Advances in Bioelectrochemistry Volume 1

Frank N. Crespilho
Editor

Advances in Bioelectrochemistry Volume 1

Surface, Electron Transfer and Techniques

Springer

Editor
Frank N. Crespilho
University of Sao Paulo
São Carlos, São Paulo, Brazil

ISBN 978-3-030-94987-7 ISBN 978-3-030-94988-4 (eBook)
https://doi.org/10.1007/978-3-030-94988-4

This Springer imprint is published by the registered company Springer Nature Switzerland AG
The registered company address is: Gewerbestrasse 11, 6330 Cham, Switzerland

Contents

Computer Simulation: Biomolecules on Surfaces

Filipe Camargo Dalmatti Alves Lima, Luana Sucupira Pedroza, Rodrigo Garcia Amorim, and Wanderlã Luis Scopel

1 Introduction

Most chemical and physical properties of atomic systems are related to electronic properties. Those properties are still one of the most studied topics since they can behave differently among molecules, solids, interfaces, and their interactions. In the last decades, experimental techniques became more sophisticated, being able to reach the electronic structure and to provide unique fingerprints. One of the hot topics studied is surface science, due to its large area leading to a potential for applications, such as biomolecule identification. Due to the experimental challenges, one way to have better comprehension of quantum effects at surfaces is using theoretical modeling. In particular, Density Functional Theory (DFT) is a powerful tool to study the electronic structure, electron transfer, binding energies, electron transport, and currents, being able to provide further insights about these systems.

Surface modeling is a many-body problem that involves several atoms and electrons. In quantum mechanics, this type of problem does not have an analytical solution. Just a few simple problems have an analytic solution: a free particle in a box, quantum oscillator, and hydrogen atom. The first approximation that can be done to

F. C. D. A. Lima (✉)
Federal Institute of Education, Science, and Technology of São Paulo (IFSP),
Matão, SP 15991-502, Brazil
e-mail: fdlima@ifsp.edu.br

L. S. Pedroza
CCNH - Center for Natural Sciences and Humanities, Federal University of ABC (UFABC),
Santo André, SP 09210-580, Brazil

R. G. Amorim
Physics Department, ICEx, Fluminense Federal University (UFF),
Volta Redonda, RJ 27213-145, Brazil

W. L. Scopel
Physics Department, Federal University of Espírito Santo, Vitória, ES 29075-910, Brazil

F. N. Crespilho (ed.), *Advances in Bioelectrochemistry Volume 1*,
https://doi.org/10.1007/978-3-030-94988-4_1

simplify the problem is using the Born-Oppenheimer approximation: the electrons have different time scale compared with the nuclei, due to their mass difference ($m_N \gg m_e$). Thus, it is possible to consider the nuclei fixed to solve just the electronic hamiltonian. Although this problem becomes simpler to deal with, it is still unsolvable. There are different methods to treat this problem based on wavefunction, that depends on 3N variables without considering the spin. An example of a wavefunction method is Hartree-Fock [60]. For large systems, this method is extremely costly. One way to overcome this very hard problem is using the charge density instead, which is more efficient to treat compared to wavefunction's method. DFT was proposed by Hohenberg and Kohn in 1964 [26], which drastically changes how the results from the quantum mechanics theory are obtained. In 1965, Kohn and Sham [28] proposed mapping of N-body problem in N problems of one body. For that, they proposed the self-consistent equations and defined the KS potential that has the external potential, Hartree, and the exchange-correlation potential, which are known as Kohn-Sham Scheme (KS) for the DFT (KS-DFT).

The KS wavefunctions can be described using different mathematical approaches. There are many computational codes available that have a particular description, having some characteristics with advantages and disadvantages. Computational codes such as Gaussian [17] or ORCA [43] employ wavefunctions based on Pople's Gaussian orbitals in real space, being very effective to describe molecular structures, however struggles to study bulk materials or molecule/material interfaces. Quantum ESPRESSO [14, 20] and VASP [29] on the other hand, describe the wavefunctions as planewave basis sets in reciprocal space, using pseudopotentials [62] or projector augmented wave (PAW) [6] approximations to describe the core region of the atoms. While reciprocal space methods are well suited for bulk materials and interfaces, their computational cost is hugely increased in comparison to real space methods. There are also computational codes such as SIESTA [18] that employs numerical localized basis sets, pseudopotentials and linear-scaling algorithms in real space, allowing investigations of molecules, bulk material and interfaces. In the end, the computational packages choice is based on which property of interest, or the type of material, is investigated. There are state-of-art theories and computer implementations that are exclusive to it package.

Another critical point to establish the DFT accuracy, precision and reliability is the exchange correlation functional (E_{xc}) employed for the system. The most common used in the electronic structure calculations are the Local Density Approximation (LDA) [48], generalized gradient approximations (GGA) such as PBE [49] or PBEsol [50], and hybrid functionals such as B3LYP [5] or HSE [25]. If a system has non-bonded interactions, van der Waals corrections should be included [21]. As discussed before, there is not an unique choice since it depends on the system model, the analysis, the computational code, convergence tests and the computational resources available. There are many studies that provide E_{xc} benchmarks, for example Su et al. [59].

The analysis of the electronic structure calculations allows to understand the stability, bonding, potential energy surface, density of states (DOS), band structure, electronic transport and others properties for different systems. The highlight properties could guide and/or support the experimental results, leading to the understanding of fundamental atomic aspects.

The KS wavefunctions are the eigenvectors obtained from the KS equation, associated with the energy levels/bands, that express the probability to find an electron in a given region of the space. They can be used to understand where the electrons are localized in the molecule/solid and how they can behave. In bioelectrochemistry, this analysis can help to understand which atoms can possibly participate on the electron transfer process between the surface and molecule.

The DOS accounts the electronic states that are found in a particular energy, or in an energy window. It can be related to physical or chemical effects, such as surface interactions, reaction centers, defects, electron entrapment and others. Also, it can be used to reproduce experimental techniques such as scanning tunneling microscopy (STM). The total DOS for a given system is difficult to interpret since there are many states involved. In this scenario, projected DOS for a given atom and/or orbitals, or atomic groups are employed instead. This helps to understand the specific role for the components of the system.

The electronic transport is a powerful model to understand microscopically what is happening in a electrochemistry process and nanodevices such as biosensors, etc. The concept of the electronic transport is different of canonical electronic calculation in equilibrium. In this case, two electron reservoir or electrodes are considered, and the central region is called the scattering region. There is one electrode in the left side and other in the right of the scattering region. The system is considered semi-infinity in both directions [7, 54]. The basic quantity calculated is the transmission function, that represents the probability of one electron, from the left side, to reach the right electrode passing through the scattering region. The transmission relevance is due to the possibility to obtain the electronic conductance and current. In the case of biosensors, one can calculate the transmission for the bare system and to obtain the zero bias conductance/resistance. For the molecule chemio/phisiosorbed in the device, it is possible to calculate the new transmission, where the charge transfer between the molecule and device will play the main role. Changes in the conductivity/resistance (sensitivity) can be a powerful tool to identify electrical signature of biomolecules in such systems.

The molecule adsorption can be understood by the total charge density that is given by the square of KS wavefunctions. The charge density gives the electron probability and possibly quantifies the referred property. For example, in a pristine material, the charge density can characterize the bond nature. If the system is composed by a surface and a molecule, it is possible to evaluate the electronic charge reorganization through the charge density difference. In this case, we consider the difference between the whole system (surface + molecule) and isolated counterparts. From the mathematical point of view, it is possible to write:

$$\Delta\rho(\vec{r}) = \rho_{surface+molecule}(\vec{r}) - (\rho_{surface}(\vec{r}) + \rho_{molecule}(\vec{r})), \tag{1}$$

where the regions can be locally analyzed: in the case of $\Delta\rho(\vec{r}) > 0$, that region decreases the charge, and for $\Delta\rho(\vec{r}) < 0$, the charge increases when compared to the isolated parts. This quantity gives how the charge is redistributed in the entire system compared with the isolated ones. The charge redistribution is an important way to quantify the charge transfer between molecule and surface.

Richard Bader [4] has proposed an intuitive scheme for visualization of atoms in molecules focusing on the charge density. Bader has used zero flux surfaces to separate the atomic charge, where the electronic charge density has a minimum value perpendicularly to the surface. Typically in molecular systems, the charge density reaches a minimum between neighbour atoms, which is a natural place to separate them from each other. Therefore, Bader considered the electronic charge density enclosed within this volume as a good approximation for the total electronic atom charge. Thus, obtaining the partial atomic charges. This quantity is interesting to investigate local reorganization, reduction and oxidation states in the bioelectrochemistry context. For computational Bader charge analysis, there is a open source code developed by Henkelman Group [24, 66].

In the following sections, we illustrate the state of the art of theoretical models of biomolecules interface with surface or nanodevices. These systems were treated within quantum mechanics calculations, using the analysis tools summarized here.

2 Electron Transfer of Peptide SAMs Modified with Ferrocene

The heterogeneous electron transfer (ET) involves a molecule/electrode interface that is composed by a donor-acceptor system: the electroactive group is connected to the electrode through a spacer molecule, which can be a polypeptide, alkanethiol, lipid, sugar, and others. In particular, ferrocene (Fc) modified self-assembled monolayers (Fc-SAMs) have been considered as excellent models for the study of ET [13, 40, 55]. One interesting study has observed higher ET rate in peptides compared to thiol ligand, both functionalized with Fc [45]. Despite the progress in the area, there is a dichotomy regarding the ET mechanism [38]: from one side they believe that the electron hopping through amide groups whereas the electron moves within sites from the donor to the acceptor; on other hand, electron tunneling where the electron moves from the donor to acceptor without any intermediate state.

Based on this paradigm, the community carried out new experiments [13, 27, 37, 39, 42, 47, 51, 55, 61, 65], emerging three possible types of ET mechanisms: molecular mobility, electron tunneling, and hopping, as well as combinations of those [8, 39, 61].

From the theoretical perspective, Lima et al. [33] carried on DFT calculations of several models to investigate each component in the system. For calculations, the authors used DFT and GGA exchange-correlation functional, other computational details can be found elsewhere [33]. These tests investigated the role of the space molecule glycylcistamine (Gly-CSA), the electroactive center Fc and the electrode modeled by the Au(111) surface. It is worth to observe that the electronic states that belongs to the Gly-CSA are always below the Fermi energy, even during simulation considering Gly-CSA@Au(111). On the other hand, the Fc electronic states are always located close to the Fermi energy, regardless the configuration it was investigated. In summary, Fig. 1 shows a gold slab with the Fc-SAMs with the organic solvent environment. The DOS analysis in this case indicates the electronics states closer to the Fermi Energy (E_F) are the thiol bond (S-Au), then the Fc group (H_M) without any intermediate state from the spacer (H_m) or the solvent. In terms of the energetics, while the Fc peak is placed at −0.3 eV, the electronic contributions for the molecular spacer are located about −1.5 eV below the E_F. Although DFT simulations presented do not show the dynamics of ET, the results demonstrate the electronic pathways. In this case, it is possible to describe the ET involving the Fc group and the gold electrode, favoring the electron tunneling mechanism. The DOS analysis combined with different building models contributed to elucidate the charge mechanism discussion in Fc-SAMs.

3 Two-Dimensional (2D) Systems

In 2004, graphene mechanical exfoliation was assembled by Geim and Novoselov [44], opening new perspectives in 2D materials and its applications. Graphene is a lightweight, atom-thick, zero-gap semiconductor, it has a large Young's Modulus (about 1 TPa) that is flexible perpendicularly to its basal plane. These unique properties in a single material make it special among others. One of the promising applications of graphene is using as a fast biomolecule sensor, i.e. utilize it as a solid-state nanodevice for electric identification. Graphene paves the way for 2D materials [16], and nowadays more than 700 different structures are known. It is possible to highlight some of 2D materials such as: h-BN [3], silicene [2], germanene [34], borophene [64], phosphorene [10], transition metal chalcogenides (TMDs) [12], Janus [67], also combinations of them (hybrid materials) [22, 56, 58]. Considering the electrical properties, the 2D materials can be metallic, semi-metallic, semiconductor and insulator. The material properties control allows different technological applications e.g.: biosensors, next generation batteries and nanodevices.

For biomolecule electric sensing, there are three possible mechanisms discussed by Heerema [23] and Diventra [11], where the schematics are depicted in Fig. 2: (a) on top, (b) on edge or nanogap and (c) nanopore device, respectively.

On top mechanism uses a large 2D surface area for device biomolecule identification. In this mechanism, the molecule is lying on the surface as could bind strongly (chemisorbed) or weakly (physisorbed), depending on the molecule nature. In general, the device has a bare resistance, and each molecule has different electron transfer. This charge transfer and perturbation leads to changes in the device resistances, thus the sensitivities can be characterized. A typical biomolecular signature is the Fano resonances [41], and the challenge is to distinguish a molecule among others based on their signal. Another important aspect to be considered for the nanodevices consists to detect the biomolecules with binding energies smaller than 1.0 eV. The

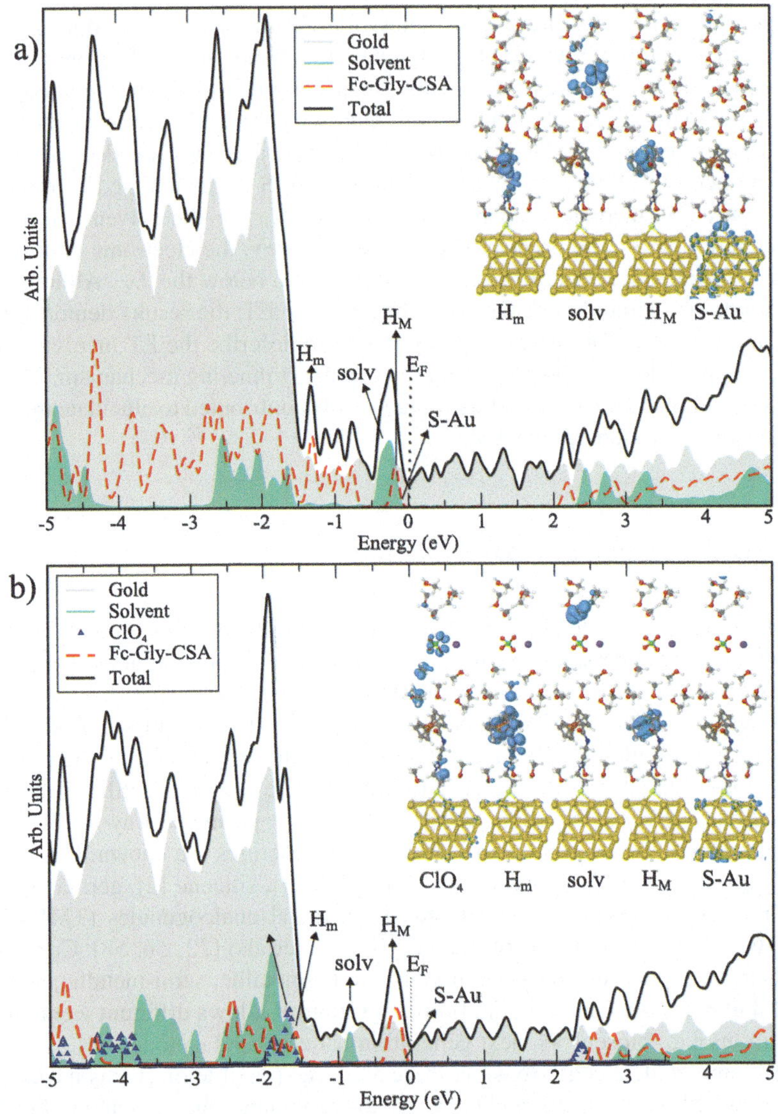

Fig. 1 Total DOS (black line) and projected contributions on the Au substrate (shaded area) and Fc-Gly-CSA molecule (red line) for the molecule/surface interface: **a** including methanol solvent (cyan area); **b** including sodium perchlorate (blue line) and methanol solvent. Zero energy reference is set to the Fermi energy (E_F) of the system. Insets display representative single particle orbitals of the interface. Reprinted (adapted) with permission from Ref. [33]. Copyright 2014, American Chemical Society

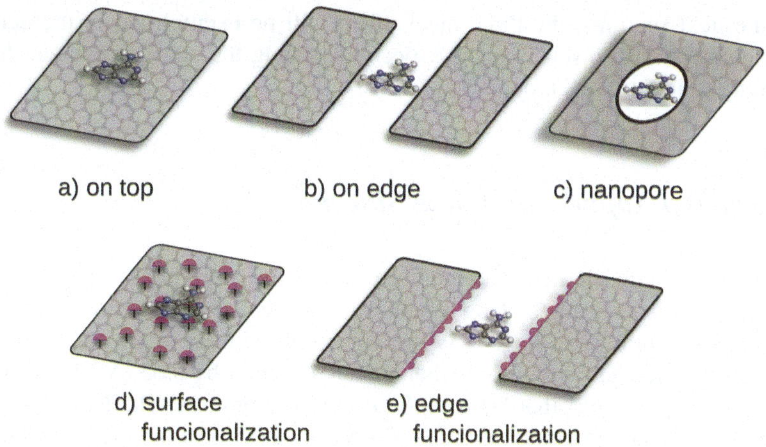

Fig. 2 Systems based with graphene, with possible site interactions: **a** on top; **b** on edge or nanogap; **c** nanopore; **d** surface functionalization; **e** edge functionalization

translocation time is given by $\tau \sim \exp(\frac{-E_b}{K_B T})$, where E_b is the binding energy, K_B is the Boltzmann constant and T is temperature, that characterize the molecule resident time in the device. Large binding energies could lead to disposable nanodevice, once the translocation time could be very large. For the case of very small molecule binding, it is possible to consider the surface functionalization as it is shown in Fig. 2d. The functionalization should be an atom or a molecule that acts an anchor to interact with the biomolecule.

Figure 2b is shown on edge or nanogap device based on tunneling mechanism. For a bare system, the transmission is zero per construction due to the large distance between the sides. When the molecule is placed on the nanogap, it can work as a bridge between the two sides (electrodes) and the electron could tunnel from the left to the right side. The conductance typical signal is about 10^{-10} G_0 and this fact could be a experimental challenge due to the signal-to-noise ratio. A possible way to minimize this problem is the edge functionalization with different atoms as it is shown in Fig. 2e. The functionalization can increase the conductance signal and signal-to-noise ratio. This fact could improve the tunneling transmittance signal for different biomolecule, that also depends on the coupling.

The nanopore is the third and more promising biomolecule nanodevice as it is shown in Fig. 2c. In this case, there are a nanodevice with 10–20 Å pore diameter. This pore size allows the biomolecule passing through it, changing the electronic characteristics of system. In this mechanism the bare device has a resistance, when the molecule is inside the pore it changes, which is similar to the on top. In fact, the molecule can modulate the electronic current leading to its identification. Nowadays, there are technics to control the nanopore size, shape and its functionalization, fundamental for this type of applications.

In the next sections, the three mechanism will be exemplified in the detection of nucleobases, also discussing the aspects regarding the change transfer, that is a possibility to identify each one.

3.1 On Top Application Using Silicene

Silicene is a silicon 2D material, similar to graphene, with honeycomb lattice that was proposed theoretically by Ciraci et al. [9]. The electronic structure is a semimetal system with a Dirac cone in K-K'-symmetry point, similar to graphene. However, silicene is buckled and more stable than planar structure by 0.05 eV [2]. This fact is due to $sp^2 - sp^3$ bonds, that leads to the referred roughness. The lattice parameter for buckled (planar) is about 4.002 (3.995) Å with 0.55 Å of height, using DFT and GGA + van der Waals correction for exchange correlation functional as implemented in SIESTA code. Further details can be find elsewhere [2]. Considering the transport point of view, it was build a supercell that are coupling two electrodes in left/right sides. The transmission gives a V-shape curve, in good agreement to Dirac materials. Biomolecules such as: adenine (A), cytosine (C), guanine (G) and thymine (T), were adsorbed on the silicene surface and distinct orientations were fully relaxed. With the most stable geometries of each nucleobase were calculated the binding energy and the distance between the molecule and device. For A and T, the distance was about 3.0 Å and the binding energy close to −0.6 eV. However, for C (G) the distances were about 2.0 Å and the binding energy were −0.88 (−0.77 eV), respectively. The strong binding for C and G is related to single oxygen atom presented in each nucleobase. This fact makes the molecule tilt, where the Si-O atoms are minimal distance between the surface and biomolecule. For A and T the molecules are planar, where the A does not have oxygen atoms, and T has two (one in each molecule side). For the T molecule, there are oxygens competition and the molecule stays quasi planar. Based on these results, it is possible to classify the nucleobase in two groups: (I) weak binding (A and T) molecules and (II) strong binding (C and G).

Figure 3 shows the difference of charge density calculated using Eq. 1. For the group I, it is noted a charge repulsion between the biomolecule and the silicene device. For A and T (Fig. 3a, d), the negative difference (blue color) is between the bottom of the molecule and the surface. In another hand, Fig. 3b, c represent the group II, where the molecules (C and G) are stronger in terms of binding energy, leading to an attractive charge redistribution. In the inset, can be noted a negative charge concentration between Si-O and a positive charge on Si atoms, confirming this assumption. These results confirm the binding energy previously discussed.

Next, we analyse the electronic transport by the transmission function. For C and G molecule, it was noted a decreasing in the transmission at +1.0 eV, compared with pristine device and also A and T. A similar result is also noted from −1.0 eV to E_F, where the group II decreases the transmission compared to the pristine one. Fano resonances were observed for energies smaller than −1.0 eV, typical from molecular signatures. For the inset upper (lower), two energies were chosen for the

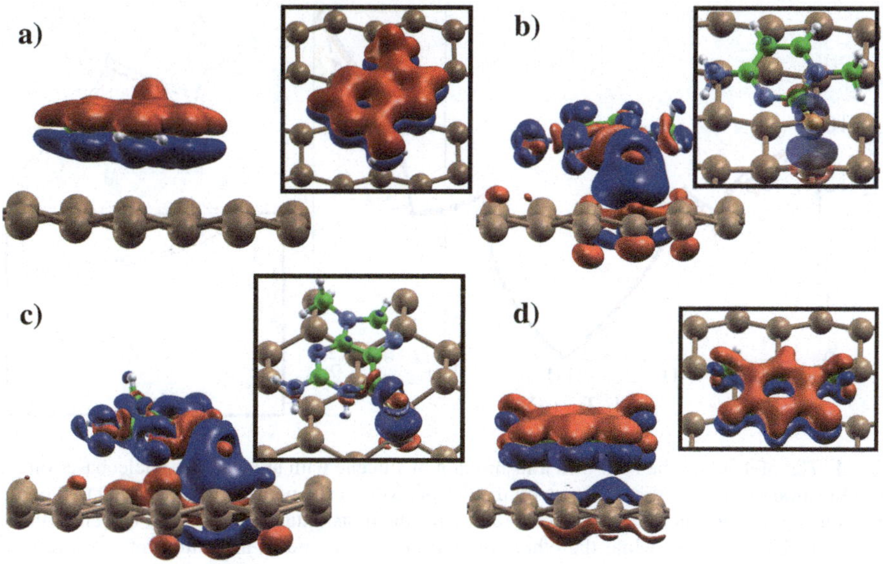

Fig. 3 The change in electronic charge density is plotted, calculated as the difference between the charge density of the total system (silicene + nucleobase) and that of each constituent part (silicene, nucleobase) separately. Blue color indicates a negative difference in the charge density; red color indicates a positive difference in the charge density. The main panels show isosurfaces for a value of 0.001 electron $\times$ bohr^{-3} while the insets show the same charge density difference data plotted for a larger isosurface value of 0.03 electron $\times$ bohr^{-3} and viewed from a different perspective. **a** Adenine; **b** cytosine; **c** guanine; **d** thymine. Reprinted (adapted) with permission from Ref. [2]. Copyright 2015, IOP Publishing, Ltd.

analysis of the changes in the transmission. In both cases, the molecules transmission are different than the pristine one. The transmission changes observed leading to the modification in the resistance/conductance, important aspect to molecule identification.

For electrical nucleobase identification were chosen two energies values +1.0 eV and −1.26 eV (upper and lower right side panel on Fig. 4), at the Fano resonances. One way to identify changes in the conductance/resistence is through the sensitivity, that is given by:

$$S[\%] = \left(\frac{g_0 - g_i}{g_0}\right) \times 100, \tag{2}$$

where g_0 is the pristine conductance and g_i is the nucleobase conductance (for i = A, C, G and T). Figure 5 shows the sensitivity, where for a gate voltage of −1.26 eV is noted a sensitivity up to 20% to A and about 3% for T. The nucleobase C and G has similar sensitivities (around 10%). However, considering the gate voltage of +1.0 eV it is possible to identify C and G with a good resolution and A and T are indistinguishable. This device can identify each nucleobase using two gate voltages through sensitivities, as well as with translocation time with a good resolution.

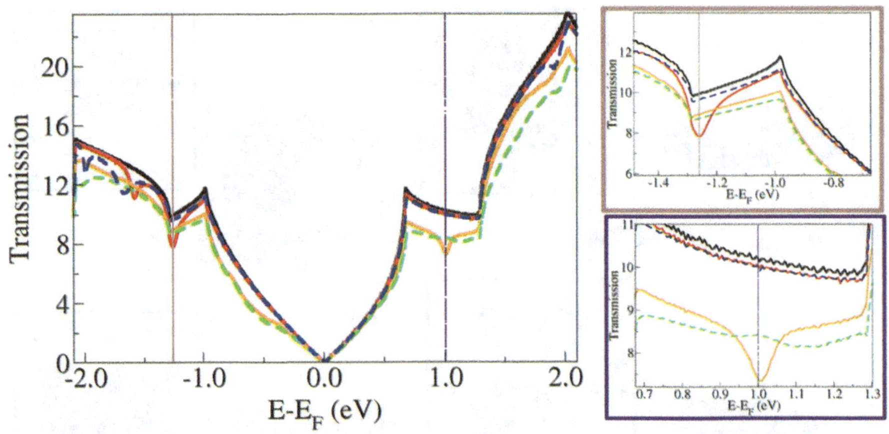

Fig. 4 The plot shows the zero-bias transmission of silicene with and without nucleobases on top as a function of the electron energy, with the Fermi level for the whole system aligned to 0 on the horizontal axis. The curve in black color represents the transmittance for pristine silicene without any nucleobases present while the other colors (red, orange, green, and blue) refer respectively to the transmission for each nucleobase (A, C, G, and T, in that precise order) physisorbed or chemisorbed on top of silicene. Reprinted (adapted) with permission from Ref. [2]. Copyright 2015, IOP Publishing, Ltd.

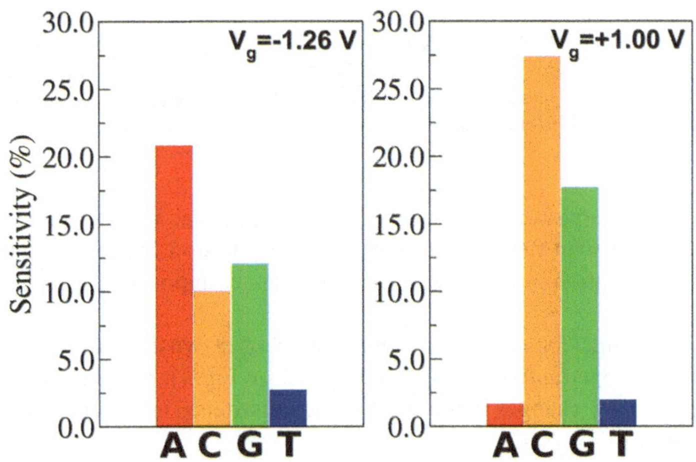

Fig. 5 For two different gate voltages (left panel: −1.26 V; right panel: +1.00 V) we plot the calculated sensitivity of the hypothetical silicene sequencing device with respect to the four different nucleobases A, C, G, and T). Reprinted (adapted) with permission from Ref. [2]. Copyright 2015, IOP Publishing, Ltd.

3.2 *Nanogap Application with Graphene*

Nanogap or edge device consists of surface material with a empty space between two regions. From the experimental point of view, Postma [52] proposed a graphene-based device for reading the transverse conductance of biomolecules, in particular nucleobases. In the proposed setup, it was considered the electrodes as the same material, demonstrating the nucleobase electrical detection with single-base resolution. Theoretically, a graphene nanogap device was investigated by Jariyanee Prasongkit et al. [53], where the tunneling transmission were studied, and the authors showed very small electrical conductance (in the order of 10^{-10} G_0). The main question to be addressed consists on how can we modify the nanogap device to improve the transport properties. To answer this issue, the nanogap functionalization will be addressed to electrically detect biomolecules with better conductance signal.

The results shown here were obtained using DFT calculations with GGA exchange-correlation functional, further details can be found elsewhere [1]. Graphene nanogap with H- or N-edge functionalization were considered. The devices were built and full ionic relaxation were performed. Thus, the shortest distances between the two sides are about 13.85 Å. The A, C, G, and T nucleobases were placed as a linker between the two sites. There are two molecules that have two rings, namely group I (A and G), and two with one (C and T) as group II. For group II, it was expected that the small size, in comparison with group I, can lead to a weaker coupling. The nucleobases were relaxed inside the nanogap and for the group I were found distances smaller than group II for both devices.

Figure 6 shows the total density of charge for H- (left side), N-functionalized device (right side), and each row represents one nucleobase. The nitrogen is characterized by a high electron concentration (blue color) in comparison to the hydrogen one. Firstly, were observed a weaker coupling for the biomolecules of group II in contrast with the group I, as expected. In another hand, analyzing A as an example and comparing to a different device, were noted a strong interaction in a N-functionalized one. Figure 6 shows an inset in red for weak coupling (H-functionalized device) and N-functionalized the green/yellow. For A and C in H-functionalized devices, it is noted the smaller nucleobase presented a weaker interaction. This fact is also true for N-functionalized devices. However, for C in N-functionalized devices were observed a better electronic coupling in comparison with H-functionalized devices. These results show that group I has stronger coupling and the N-functionalized device is better for biomolecule detection in comparison to H one.

Figure 7a–b show the zero bias transmission for H-passivation (left panel) and N-passivation (right panel), respectively. Firstly, it was investigated the four nucleobases at each nanogap device. At Fermi energy, the transmission hierarchy was $G > A > C > T$ for both devices (-H and -N) and the main difference is the transmission order of magnitude. For N-functionalization it was noted a conductance improvement about $10^4 - 10^5$ G_0. For all cases, there are resonances for a certain energy value due to the nucleobases. For nucleobase A, there is a peak at -1.0 eV for H-functionalized nanogap. Secondly, for N-functionalized devices, the previous peak moves towards

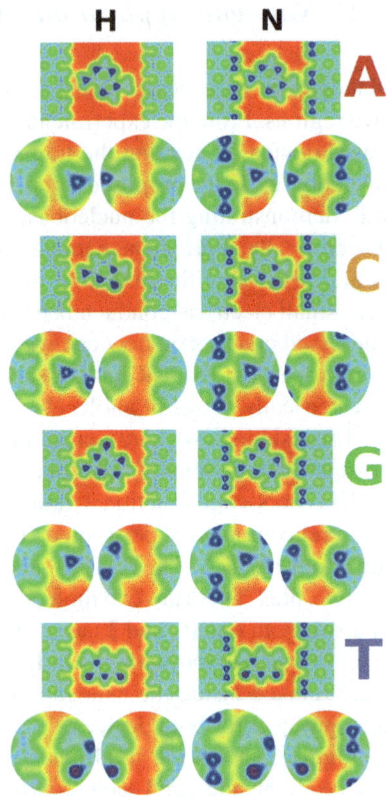

Fig. 6 Charge density for a graphene nanogap with nucleobase in between where left handside panel is for hydrogen, and right is for nitrogen functionalization. The circular insets show zoomed in regions at the frontier between molecule and edge. The colors represent the charge density is given in linear scale going from 0.0 (red) to 1.1 (violet) and $e/Bohr^3$. Reprinted (adapted) with permission from Ref. [1]. Copyright 2016 American Chemical Society

Fermi energy. This fact indicates that further device has a better coupling, as it was previously demonstrated in the total charge density analysis. When the molecules are placed in the N-functionalized material, there was noted a peak shift toward the Fermi energy.

For the nanogap case, the sensitivity characterization is difficult to evaluate due to zero transmission for the bare system (without the molecule). For the analysis of the changes in the conductance without device reference, it was considered a target molecule. In this case, it was chosen a molecule with the smaller conductance signal, which was the nucleobase T at the H-functionalized device. In this way, all nucleobase conductance were normalized by T, and the sensitivity definition is given by the following:

$$S[\%] = \left(\frac{g_i}{g_{ref}}\right) \times 100, \tag{3}$$

where g_i represents the nucleobase conductance, and g_{ref} the smaller conductance case (T in H-functionalized device). Figure 8 hashed histogram shows the H-functionalized, where the biggest sensitivity is given by G, followed by

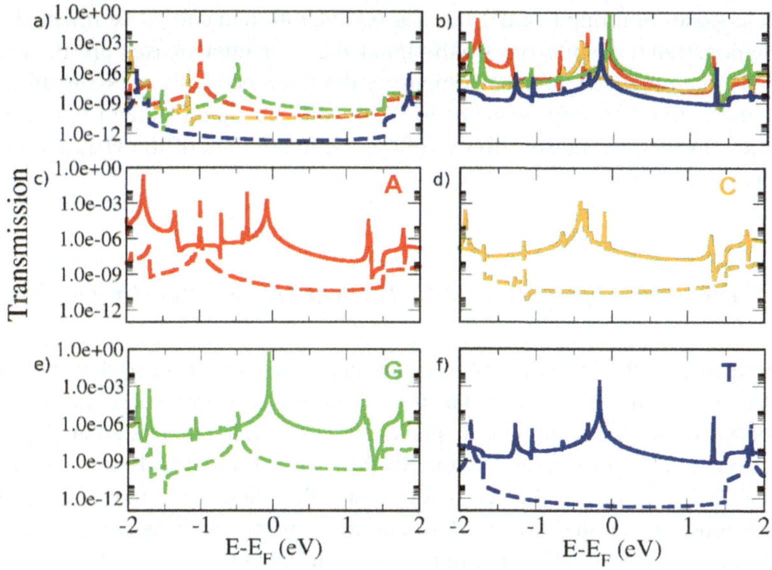

Fig. 7 Zero-bias transmission as a function of energy: graphene nanogap functionalized with **a** Hydrogen (dashed), and **b** Nitrogen atoms (solid) for all nucleobases; **c–f** transmittance comparison between H- and N-terminated edges of graphene nanogap with different nucleobases in between—A, C, G and T. Reprinted (adapted) with permission from Ref. [1]. Copyright 2016 American Chemical Society

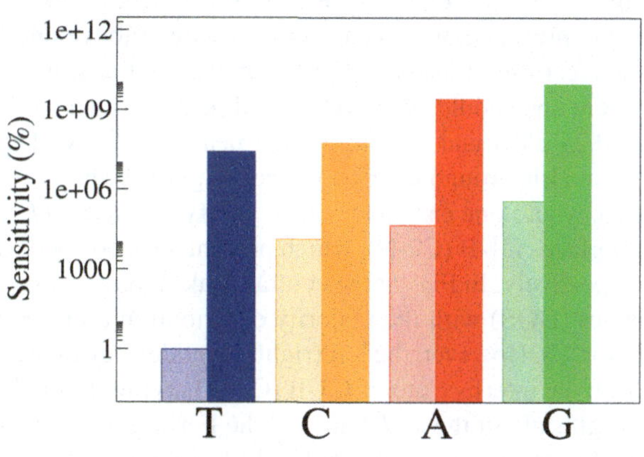

Fig. 8 Sensitivity at Fermi level of four nucleobases for H- (hashed) and N-functionalized (solid) nanogaps. Thymine in the H-terminated nanogap has the smallest conductance and was taken as a reference. Reprinted (adapted) with permission from Ref. [1]. Copyright 2016 American Chemical Society

A—nucleobases of group I, and C has a good resolution in comparison with T. Here, were demonstrated despite the small-signal the four nucleobases are distinguishable in H-functionalized device. Comparing the reference with an N-functionalized device, were noted an improvement in the sensitivity by more than five orders of magnitude, as well as the sensitivity follow the magnitude of the H-functionalized device ($G > A > C$).

3.3 Nanopore Application with Hybrid Graphene-BN Device

Nanopore inside of 2D materials has emerged as a great potential to be used as DNA sequencing due to its individual-base resolution. The applicability of solid-state nanopores for the detection and analysis of biomolecules is dependent on achieving accurate control of size and shape. In this context, recent advances in size and shape control have been achieved, using controlled dielectric breakdown [30] and electrochemical reaction (ECR) [15]. Encouraged by the experimental synthesis successfully of nanopores embedded in the 2D materials [15], Souza et al. [56], have been proposed a novel device based on lateral heterostructure graphene-BN.

For modeling the proposed systems herein, it was performed DFT calculation with GGA exchange-correlation functional, more details can be found somewhere [56]. The investigated setup has two electrodes (left and right) and a scattering region with a nanopore. The graphene stripe has a width of 16.92 Å embedded in the h-BN and the pore diameter is 12.5Å (see Fig. 9e). One important aspect of the device is the presence of two carbon nanowires (above and below the pore). Focusing on the electronic transport, Fig. 9a shows the transmission spectra for the bare device, where one should verify two broad peaks around the Fermi level in the energy range (±0.25 eV)) and a well-defined peak at 0.9 eV. The transmission is actually not symmetric around the Fermi level. Figure 9b shows the vertical dashed lines (±0.18 eV), where one can clearly note an asymmetric $T(E)$, as a result of the two distinct interfaces C1-B (C2-N), which present, electron acceptor (electron donor) character, respectively. In Fig. 9c the central peak (orange color) represents the total density of states (DOS) with the majority contribution ascribed to central carbon atoms (blue color, C). However, the left (right) shoulders are associated with the hybridized states from the interface atoms C1-B (C2-N), as highlighted in Fig. 9d. In order to get some insights about the $T(E)$ around the Fermi level, it was analyzed the local current in the scattering region at ±0.18 eV. For the negative energy value (Fig. 9e), the local current is located in the lower carbon wire, while for the positive one (Fig. 6f) it is in the upper, as expected from the projected density of states (PDOS) (see Fig. 9c–d). The results reveal for the proposed device an important functionality, once it is possible to choose the electrons pathway switching the applied gate voltage.

The nucleobases were placed inside the nanopore and performed full relaxations. With the most stable configurations (Fig. 10a–d), the interaction strength between nanodevice and molecule were examined through the binding energy. Figure 10e shows binding energy for the four nucleotides such as: deoxyadenosine monophos-

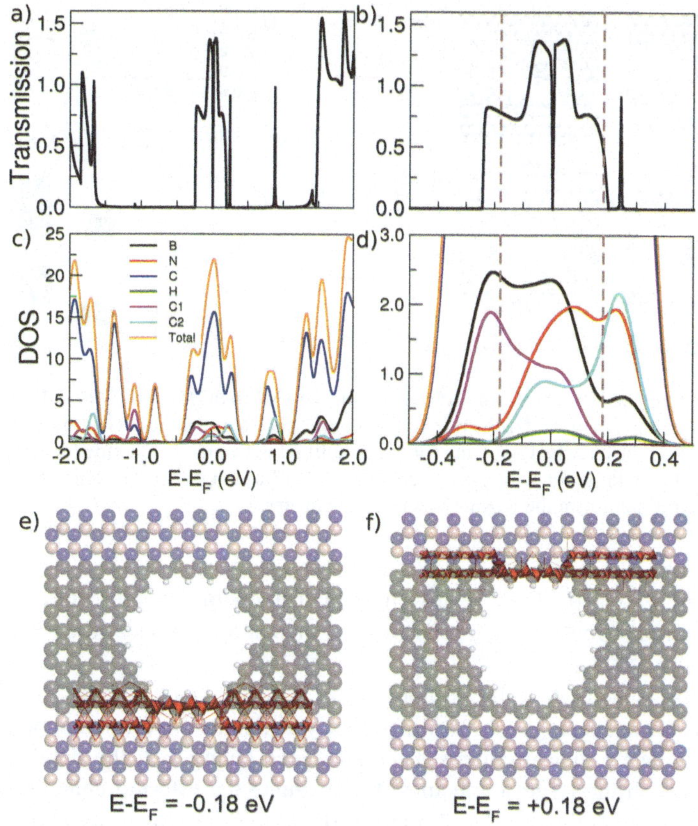

Fig. 9 **a** Zero-bias transmission spectra as a function of energy for the device without any molecule present in the pore; **b** zoom-in around Fermi level; **c** the projected density of states (PDOS) as a function of energy; **d** zoom-in of the PDOS around Fermi level; the local current density **e** for −0.18 eV and **f** for +0.18 eV. Reproduced from Ref. [56] with permission from the Royal Society of Chemistry

phate (dAMP), deoxyguanosine monophosphate (dGMP), deoxycytidine monophosphate (dCMP) and deoxythymidine (dTMP). In particular, the dCMP is 0.65 eV and dAMP is 0.13 eV, which is higher than corresponding binding energies on graphene and MoS_2 surfaces [31, 32]. Then, based on the binding energy values, one can infer that dCMP will have the biggest resident time while dAMP will have the smallest one. These results suggest that the device might also allow to identify individually different target molecules due to their specific residence times into the nanopore.

Figure 10f shows the transmission spectrum of the bare device (black dashed line) and device with four nucleotides. These results reveal that one can distinguish the nucleotides based on their transmission signature around the Fermi level. In general, one can observe that $T(E)$ decreases for all nucleotides relative to the bare nanopore system. However, for one case (i.e., dGMP), there is an electron channel opening for

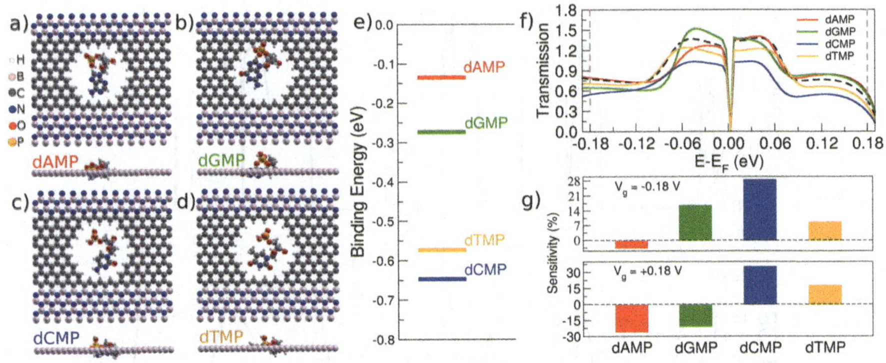

Fig. 10 Fully relaxed structures for each nucleotide **a** dAMP, **b** dGMP, **c** dCMP, and **d** dTMP inside the pore; **e** binding energies; **f** zero-bias transmission spectra in the vicinity of the Fermi level for the empty pore (black dashed line) and for the pore filled with each nucleotide (colored full lines); **g** sensitivity histograms for two specific energies (−0.18 and +0.18 eV). Reproduced from Ref. [56] with permission from the Royal Society of Chemistry

the range of −0.06 to 0.0 eV relative to the Fermi energy, which could be attributed to a charge redistribution ($\Delta\rho > 0$) close to the carbon atoms at the left side of the device's pore [57].

The main issue concerning a biomolecule sensor is its capability of identifying each molecule according to a specific property. From the experimental point of view, it can be done by measurements of conductance changes in the device due to its interaction with the target molecules. For evaluate the referred changes were used the sensitivity as defined in Eq. 2. Figure 10g shows the sensitivity at gate voltages $V_g = \pm 0.18\ V$, for the device with each nucleotide at its most stable configuration. Even though, one could present the sensitivity for fully energy spectra, herein, a specific gate voltages were chosen, as at these energies higher contributions of C1 (C2) atoms to the total DOS are noted (Fig. 9c–d). For V_g negative (positive) we noted −3.6%, 16.6%, 29.0%, and 8.3% (−25.2%, −19.9%, 36.2%, and 17.9%) for dAMP, dGMP, dCMP, and dTMP, respectively. These results indicate that device possesses a considerably sensitive resolution to distinguish each nucleotide electrically.

Figure 11 shows the local current for each nucleotide at $V_G = -0.18$ V, where the intensity is represented by the size of the arrows. Combining the sensitivity (Fig. 10g) with the electrical local current of bare device (Fig. 9e), one notes that the dCMP (dAMP) has the higher (lowest) sensitivity leading to bigger (smaller) local current modulation. It is also verified that the current modulation is proportional to sensitivity for dGMP and dTMP. In summary, it was demonstrated that the proposed device can modulate the local current and distinguish each nucleotide electronically.

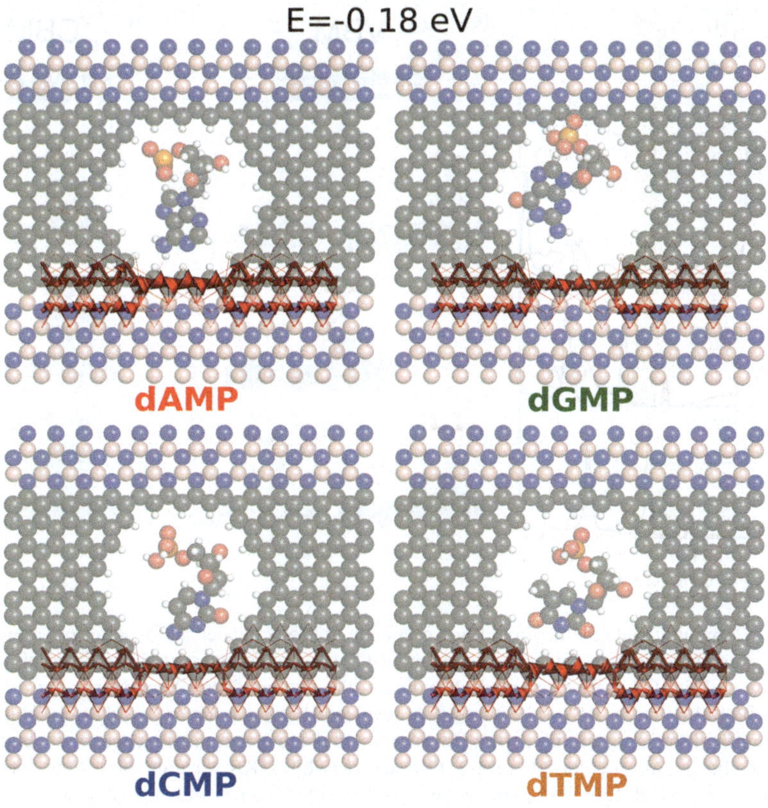

Fig. 11 Local current density modulation is shown for a gate voltage of −0.18 eV. For each nucleotide (dAMP, dGMP, dCMP and dTMP) inside the nanopore, the reference local current (Fig. 2e) is altered due to the different interaction between the molecule and the device. Reproduced from Ref. [56] with permission from the Royal Society of Chemistry

3.4 Surface Application with Graphene Functionalization

The growth or processing of graphene sheets can induce structural defects, which becomes a challenge for graphene-based nanodevice applications. To overcome this issue, the literature [19, 46, 63] has shown that these sheets can be modified with ligands to tune their properties. Surface functionalization can alter how electron transfer processes occur in graphene [16, 35], which can be applied to adjust physical and chemical properties in these systems. Macedo et al. [36] reported an experimental and theoretical study of graphene-modified with 4-carboxyphenyl (4-CP) units, indicating evidence of inhomogeneous charge distribution on the sheets. Here it was employed DFT calculations using GGA exchange-correlation functional, details can be found elsewhere [36].

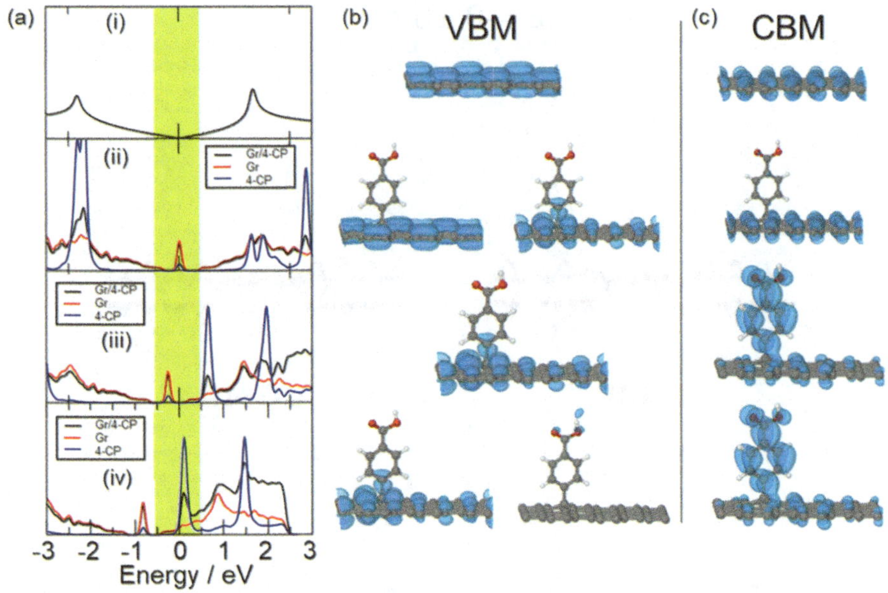

Fig. 12 Computational results of modified graphene. **a** Projected density of states per atom for (i) pristine graphene, graphene functionalized with 4-CP (ii) in a neutral system and in configurations (iii) CS-1 and (iv) CS-2. The yellow dash indicates the region of the Fermi energy between −0.5 and 0.5 eV. Kohn–Sham (KS) wavefunctions at **b** valence band maximum (VBM) and **c** conduction band minimum for cases i, ii, iii, and iv. Because ii and iv are electronically degenerate, both VBM states are shown. Atom colors: C (gray), O (red), H (white), KS wavefunction (blue spheroids). Reproduced from Ref. [36] with permission from the Royal Society of Chemistry

Figure 12 shows the projected density of states (PDOS), also the Kohn-Sham electronic wavefunctions (KS) for the valence band maximum (VBM) and the conduction band minimum (CBM). It was explored the pristine graphene (**i**) and graphene-4CP (**ii**) in a neutral system and in the configurations-1 (CS-1) (**iii**) and -2 (CS-2) (**iv**).

For the pristine graphene was observed the Dirac cone at the Fermi energy, showing π type wavefunction for the VBM and CBM. For the graphene-4CP system (**ii**), there is a hybridization between carbons atoms (graphene) with 4-CP at Fermi energy, since the total number of electrons is odd. The KS states are localized in the molecule/graphene bond in the VBM and its vicinity, while CBM states are similar in comparison to the pristine system. This study also investigates the charge injection, allowing to understand the electron transfer mechanism at the atomic level. The CS-1, denoted as the addition of one extra electron in the system. It generates a small gap and moves the central peak towards the valence band, while its VBM states are similar to the graphene-4CP in the neutral condition. Moreover, the two electrons injection (CS-2) shifts the central peak to −0.9 eV below the Fermi energy. It is worth noting that injection of the extra charge electron induces the non-uniform charge distribution in the graphene-4CP surface, as indicated by the KS VBM and

CBM wavefunctions. These results indicate an anisotropic charge distribution supporting the experimental findings. It also provided the next step, suggesting that these changes are related to the covalent anchoring in the sheet.

4 Conclusion

We presented a discussion regarding interactions between a solid surface and target biomolecules based on quantum mechanic calculations. It was performed DFT in combination with electronic transport method, analysis tools as well as applications examples. It was considered peptide SAMs and biomolecules (DNA) on the three possible detection mechanisms such as: on top, on edge, and nanopore. The on top and on edge systems were also discussed from a functionalization point of view, and their effects to improve devices electronic properties. Nowadays, the computer simulations in nanoscale systems enable a detailed atomic view of the electron transfer process between surfaces and target biomolecules. The future of computer simulations requires multidisciplinary studies between fields such as materials science, solid-state physics, biology, physical chemistry, and bioengineering among others. These advances will have meaningful impacts regarding bioelectronics, contributing to warrant the widest possible dissemination to multidisciplinary scientists, working on the cutting edge of surface science with a strong interest in applications toward nanodevices, which would inspire future experiments.

Acknowledgements The authors acknowledge the financial support of CNPq, FAPESP, FAPES, FAPERJ and INCT-INEO.

References

1. Amorim, R.G., Rocha, A.R., Scheicher, R.H.: Boosting DNA recognition sensitivity of graphene nanogaps through nitrogen edge functionalization. J. Phys. Chem. C (American Chemical Society) **120**(34), 19384–19388 (2016). https://doi.org/10.1021/acs.jpcc.6b04683. Accessed 03 Oct 2021
2. Amorim, R.G., Scheicher, R.H.: Silicene as a new potential DNA sequencing device. Nanotechnology (IOP Publishing) **26**(15), 154002 (2015). https://doi.org/10.1088/0957-4484/26/15/154002. Accessed 03 Oct 2021
3. Amorim, R.G., Zhong, X., Mukhopadhyay, S., Pandey, R., Rocha, A.R., Karna, S.P.: Strain- and electric field-induced band gap modulation in nitride nanomembranes. J. Phys. Cond. Matter **25**(19), 195801 (2013). https://doi.org/10.1088/0953-8984/25/19/195801
4. Bader, R.F.W.: Atoms in Molecules: A Quantum Theory, vol. 22. Inglês. Reprint edição. Oxford University Press, Oxford England: New York (1994)
5. Becke, A.D.: Density-functional thermochemistry. III. The role of exact exchange. J. Chem. Phys. **98**(7), 5648 (1993). ISBN: 0021-9606. https://doi.org/10.1063/1.464913
6. Blöchl, P.E., Först, C.J., Schimpl, J.: Projector augmented wave method: ab initio molecular dynamics with full wave functions. Bull. Mater. Sci. **26**(1), 33–41 (2003). https://doi.org/10.1007/BF02712785

7. Brandbyge, M., Mozos, J.-L., Ordejón, P., Taylor, J., Stokbro, K.: Density-functional method for nonequilibrium electron transport. Phys. Rev. B **65**(16), 165401 (2002). https://link.aps.org/doi/10.1103/PhysRevB.65.165401
8. Brooksby, P.A., Anderson, K.H., Downard, A.J., Abell, A.D.: Electrochemistry of ferrocenoyl beta-peptide monolayers on gold. Langmuir **26**(2), 1334–1339 (2010). http://www.ncbi.nlm.nih.gov/pubmed/19799404. https://doi.org/10.1021/La902402t
9. Cahangirov, S., Topsakal, M., Aktürk, E., Ahin, H., Ciraci, S.: Two- and one-dimensional honeycomb structures of silicon and germanium. Phys. Rev. Lett. **102**(23), 236804 (2009). https://link.aps.org/doi/10.1103/PhysRevLett.102.236804
10. Carvalho, A., Wang, M., Zhu, X., Rodin, A.S., Su, H., Castro Neto, A.H.: Phosphorene: from theory to applications. Nat. Rev. Mater. **1**(11), 1–16 (2016). https://www.nature.com/articles/natrevmats201661. Accessed 03 Oct 2021
11. Di Ventra, M., Taniguchi, M.: Decoding DNA, RNA and peptides with quantum tunnelling. Nat. Nanotechnol. **11**(2) (2016). https://www.nature.com/articles/nnano.2015.320. Accessed 30 Sept 2021
12. Dryfe, R.A.W.: 2D transition metal chalcogenides and van der Waals heterostructures: fundamental aspects of their electrochemistry. Current Opin. Electrochem. Fundamental and Theoretical Electrochemistry • Physical and Nanoelectrochemistry **13**, 119–124 (2019). https://www.sciencedirect.com/science/article/pii/S2451910318302023, Accessed 03 Oct 2021. https://doi.org/10.1016/j.coelec.2018.11.021
13. Eckermann, A.L., Feld, D.J., Shaw, J.A., Meade, T.J.: Electrochemistry of redox-active self-assembled monolayers. Coord. Chem. Rev. **254**(15–16), 1769–1802 (2010). http://www.pubmedcentral.nih.gov/articlerender.fcgi?artid=2885823. https://doi.org/10.1016/J.Ccr.2009.12.023
14. Enkovaara, J., Rostgaard, C., Mortensen, J.J.: Advanced capabilities for materials modelling with Quantum ESPRESSO. J. Phys. Cond. Matter **29**(46), 465901 (2017). http://stacks.iop.org/0953-8984/29/i=46/a=465901. https://doi.org/10.1088/1361-648X/aa8f79
15. Feng, J., Liu, K., Graf, M., Lihter, M., Bulushev, R.D., Dumcenco, D., Alexander, D.T.L., Krasnozhon, D., Vuletic, T., Kis, A., Radenovic, A.: Electrochemical reaction in single layer MoS2: nanopores opened atom by atom. Nano Lett. (American Chemical Society) **15**(5), 3431–3438 (2015). https://doi.org/10.1021/acs.nanolett.5b00768. Accessed 30 Sept 2021
16. Ferrari, A.C., Bonaccorso, F., Fal'ko, V., Novoselov, K.S., Roche, S., Bøggild, P., Borini, S., Koppens, F.H.L., Palermo, V., Pugno, N., Garrido, J.A., Sordan, R., Bianco, A., Ballerini, L., Prato, M., Lidorikis, E., Kivioja, J., Marinelli, C., Ryhönen, T., Morpurgo, A., Coleman, J.N., Nicolosi, V., Colombo, L., Fert, A., Garcia-Hernandez, M., Bachtold, A., Schneider, G.F., Guinea, F., Dekker, C., Barbone, M., Sun, Z., Galiotis, C., Grigorenko, A.N., Konstantatos, G., Kis, A., Katsnelson, M., Vandersypen, L., Loiseau, A., Morandi, V., Neumaier, D., Treossi, E., Pellegrini, V., Polini, M., Tredicucci, A., Williams, G.M., Hong, B.H., Ahn, J.-H., Kim, J.M., Zirath, H., van Wees, B.J., van der Zant, H., Occhipinti, L., Di Matteo, A., Kinloch, I.A., Seyller, T., Quesnel, E., Feng, X., Teo, K., Rupesinghe, N., Hakonen, P., Neil, S.R.T., Tannock, Q., Löfwander, T., Kinaret, J.: Science and technology roadmap for graphene, related two-dimensional crystals, and hybrid systems. Nanoscale **7**(11), 4598–4810 (2015). http://dx.doi.org/10.1039/C4NR01600A
17. Frisch, M.J., Trucks, G.W., Schlegel, H.B., Scuseria, G.E., Robb, M.A., Cheeseman, J.R., Scalmani, G., Barone, V., Mennucci, B., Petersson, G.A., Nakatsuji, H., Caricato, M., Li, X., Hratchian, H.P., Izmaylov, A.F., Bloino, J., Zheng, G., Sonnenberg, J.L., Hada, M., Ehara, M., Toyota, K., Fukuda, R., Hasegawa, J., Ishida, M., Nakajima, T., Honda, Y., Kitao, O., Nakai, H., Vreven, T., Montgomery, J.A. Jr., Peralta, J.E., Ogliaro, F., Bearpark, M., Heyd, J.J., Brothers, E., Kudin, K.N., Staroverov, V.N., Kobayashi, R., Normand, J., Raghavachari, K., Rendell, A., Burant, J.C., Iyengar, S.S., Tomasi, J., Cossi, M., Rega, N., Millam, J.M., Klene, M., Knox, J.E., Cross, J.B., Bakken, V., Adamo, C., Jaramillo, J., Gomperts, R., Stratmann, R.E., Yazyev, O., Austin, A.J., Cammi, R., Pomelli, C., Ochterski, J.W., Martin, R.L., Morokuma, K., Zakrzewski, V.G., Voth, G.A., Salvador, P., Dannenberg, J.J., Dapprich, S., Daniels, A.D., Farkas, Foresman,

J.B., Ortiz, J.V., Cioslowski, J., Fox, D.J.: Gaussian09 Revision D.01. Wallingford, CT, USA (2009)

18. Garcóa, A., Papior, N., Akhtar, A., Artacho, E., Blum, V., Bosoni, E., Brandimarte, P., Brandbyge, M., Cerdá, J.I., Corsetti, F., Cuadrado, R., Dikan, V., Ferrer, J., Gale, J., Garcóa-Fernández, P., Garcóa-Suárez, V.M., Garcóa, S., Huhs, G., Illera, S., Ko- rytár, R., Koval, P., Lebedeva, I., Lin, L., López-Tarifa, P., Mayo, S.G., Mohr, S., Ordejón, P., Postnikov, A., Pouillon, Y., Pruneda, M., Robles, R., Sánchez-Portal, D., Soler, J.M., Ullah, R., Yu, V.W., Junquera, J.: Siesta: recent developments and applications. J. Chem. Phys. (American Institute of Physics) **152**(20), 204108 (2020). https://aip.scitation.org/doi/10.1063/5.0005077. Accessed 01 Oct 2021
19. Georgakilas, V., Otyepka, M., Bourlinos, A.B., Chandra, V., Kim, N., Christian Kemp, K., Hobza, P., Zboril, R., Kim, K.S.: Functionalization of graphene: covalent and non-covalent approaches, derivatives and applications. Chem. Rev. **112**(11), 6156–6214 (2012). PMID: 23009634. https://doi.org/10.1021/cr3000412
20. Giannozzi, P., Baroni, S., Bonini, N., Calandra, M., Car, R., Cavazzoni, C., Ceresoli, D., Chiarotti, G.L., Cococcioni, M., Dabo, I., Corso, A.D., De Gironcoli, S., Fabris, S., Fratesi, G., Gebauer, R., Gerstmann, U., Gougoussis, C., Kokalj, A., Lazzeri, M., Martin-Samos, L., Marzari, N., Mauri, F., Mazzarello, R., Paolini, S., Pasquarello, A., Paulatto, L., Sbraccia, C., Scandolo, S., Sclauzero, G., Seitsonen, A.P., Smogunov, A., Umari, P., Wentzcovitch, R.M., Dal Corso, A., Fabris, S., Fratesi, G., De Gironcoli, S., Gebauer, R., Gerstmann, U., Gougoussis, C., Kokalj, A., Lazzeri, M., Martin-Samos, L., Marzari, N., Mauri, F., Mazzarello, R., Paolini, S., Pasquarello, A., Paulatto, L., Sbraccia, C., Scandolo, S., Sclauzero, G., Seitsonen, A.P., Smogunov, A., Umari, P., Wentzcovitch, R.M., Dal Corso, A., De Gironcoli, S., Fabris, S., Fratesi, G., Gebauer, R., Gerstmann, U., Gougoussis, C., Kokalj, A., Lazzeri, M., Martin-Samos, L., Marzari, N., Mauri, F., Mazzarello, R., Paolini, S., Pasquarello, A., Paulatto, L., Sbraccia, C., Scandolo, S., Sclauzero, G., Seitsonen, A.P., Smogunov, A., Umari, P., Wentzcovitch, R.M.: QUANTUM ESPRESSO: a modular and open-source software project for quantum simulations of materials. J. Phys. Cond. Matter **21**(39), 395502 (19 pp) (2009). http://www.ncbi.nlm.nih.gov/pubmed/21832390%20. http://arxiv.org/abs/0906.2569. https://doi.org/10.1088/0953-8984/21/39/395502
21. Grimme, S., Hansen, A., Brandenburg, J.G., Bannwarth, C.: Dispersion-corrected mean-field electronic structure methods. Chem. Rev. (American Chemical Society) **116**(9), 5105–5154 (2016). https://doi.org/10.1021/acs.chemrev.5b00533. Accessed 16 Feb 2021
22. Hassan, A., Macedo, L.J.A., Mattioli, I.A., Rubira, R.J.G., Constantino, C.J.L., Amorim, R.G., Lima, F.C.D.A., Crespilho, F.N.: A three component-based van der Waals surface vertically designed for biomolecular recognition enhancement. Electrochim. Acta **376**, 138025 (2021). https://www.sciencedirect.com/science/article/pii/S0013468621003157. Accessed 03 Oct 2021. https://doi.org/10.1016/j.electacta.2021.138025
23. Heerema, S.J., Dekker, C.: Graphene nanodevices for DNA sequencing. Nat. Nanotechnol. **11**(2) (2016). https://www.nature.com/articles/nnano.2015.307. Accessed 30 Sept 2021
24. Henkelman, G., Arnaldsson, A., Jónsson, H.: A fast and robust algorithm for bader decomposition of charge density. Comput. Mater. Sci. **36**(3), 354–360 (2006)
25. Heyd, J., Scuseria, G.E., Ernzerhof, M.: Hybrid functionals based on a screened Coulomb potential. J. Chem. Phys. (American Institute of Physics) **118**(18), 8207–8215 (2003). https://aip.scitation.org/doi/10.1063/1.1564060. Accessed 01 Oct 2021
26. Hohenberg, P., Kohn, W.: Inhomogeneous electron gas. Phys. Rev. **136**(3B), B864–B871 (1964). http://link.aps.org/doi/10.1103/PhysRev.136.B864
27. Kai, M., Takeda, K., Morita, T., Kimura, S.: Distance dependence of long-range electron transfer through helical peptides. J. Pept. Sci. **14**(2), 192–202 (2008). http://www.ncbi.nlm.nih.gov/pubmed/18035857. https://doi.org/10.1002/Psc.974
28. Kohn, W., Sham, L.J.: Self-consistent equations including exchange and correlation effects. Phys. Rev. **140**(4A), A1133–A1138 (1965). http://link.aps.org/doi/10.1103/PhysRev.140.A1133

29. Kresse, G., Furthmüller, J.: Efficient iterative schemes for ab initio total-energy calculations using a plane-wave basis set. Phys. Rev. B (American Physical Society) **54**(16), 11169–11186 (1996). https://link.aps.org/doi/10.1103/PhysRevB.54.11169. Accessed 01 Oct 2021
30. Kwok, H., Briggs, K., Tabard-Cossa, V.: Nanopore fabrication by controlled dielectric breakdown. PLOS ONE (Public Library of Science) **9**(3), e92880 (2014). https://journals.plos.org/plosone/article?id=10.1371/journal.pone.0092880. Accessed 30 Sept 2021
31. Lee, J.-H., Choi, Y.-K., Kim, H.-J., Scheicher, R.H., Cho, J.-H.: Physisorption of DNA nucleobases on h-BN and graphene: vdW-corrected DFT calculations. J. Phys. Chem. C **117**(26), 13435–13441 (2013)
32. Liang, L., Wei, H., Xue, Z., Shen, J.-W.: Theoretical study on the interaction of nucleotides on two-dimensional atomically thin graphene and molybdenum disulfide. FlatChem **2**, 8–14 (2017)
33. Lima, F.C.D.A., Calzolari, A., Iost, R.M., Crespilho, F.N., Petrilli, H.M., Caldas, M.J., Iost, R.M., Crespilho, F.N., Petrilli, H.M., Lima, F.C.D.A., Caldas, M.J., Iost, R.M., Crespilho, F.N., Petrilli, H.M.: Electronic structure of self-assembled monolayers modified with ferrocene on a gold surface: evidence of electron tunneling. J. Phys. Chem. C **118**(40), 23111–23116 (2014). http://pubs.acs.org/doi/abs/10.1021/jp506425c
34. Liu, N., Bo, G., Liu, Y., Xu, X., Du, Y., Dou, S.X.: Recent progress on germanene and functionalized germanene: preparation, characterizations, applications, and challenges. Small **15**(32), 1805147 (2019). https://onlinelibrary.wiley.com/doi/abs/10.1002/smll.201805147. Accessed 03 Oct 2021
35. Macedo, L.J.A., Iost, Ayaz Hassan, R.M., Balasubramanian, K., Crespilho, F.N.: Bioelectronics and interfaces using monolayer graphene. ChemElectroChem **6**(1), 31–59 (2019). https://onlinelibrary.wiley.com/doi/abs/10.1002/celc.201800934. Accessed 03 Oct 2021
36. Macedo, L.J.A., Lima, F.C.D.A., Amorim, R.G., Freitas, R.O., Yadav, A., Iost, R.M., Balasubramanian, K., Crespilho, F.N.: Interplay of non-uniform charge distribution on the electrochemical modification of graphene. Nanoscale (The Royal Society of Chemistry) **10**(31), 15048–15057 (2018). https://pubs.rsc.org/en/content/articlelanding/2018/nr/c8nr03893g. Accessed 30 Sept 2021. https://doi.org/10.1039/C8NR03893G
37. Malak, R.A., Gao, Z., Wishart, J.F., Isied, S.S.: Long-range electron transfer across peptide bridges: the transition from electron superexchange to hopping. J. Am. Chem. Soc. **126**(43), 13888–13889 (2004). http://www.ncbi.nlm.nih.gov/pubmed/15506726. https://doi.org/10.1021/Ja0401040
38. Mandal, H.S., Kraatz, H.-B.: Electron transfer across α-helical peptides: potential influence of molecular dynamics. J. Chem. Phys. **326**(1), 246–251 (2006). http://linkinghub.elsevier.com/retrieve/pii/S0301010406000498. https://doi.org/10.1016/J.Chemphys.2006.01.010
39. Mandal, H.S., Kraatz, H.-B.: Electron transfer mechanism in helical peptides. J. Phys. Chem. Lett. **3**(6), 709–713 (2012). http://pubs.acs.org/doi/abs/10.1021/jz300008s
40. MartiĆ, S., Labib, M., Shipman, P.O., Kraatz, H.-B.: Ferrocene-peptido conjugates: from synthesis to sensory applications. Dalton Trans. **40**(28), 7264–7290 (2011). http://www.ncbi.nlm.nih.gov/pubmed/21483964. https://doi.org/10.1039/C0dt01707h
41. Min, S.K., Kim, W.Y., Cho, Y., Kim, K.S.: Fast DNA sequencing with a graphene-based nanochannel device. Nat. Nanotechnol. **6**(3), 162–165 (2011). https://www.nature.com/articles/nnano.2010.283. Accessed 05 Oct 2021. https://doi.org/10.1038/nnano.2010.283
42. Morita, T., Kimura, S.: Long-range electron transfer over 4 nm governed by an inelastic hopping mechanism in self-assembled mono-layers of helical peptides. J. Am. Chem. Soc. **125**(29), 8732–8733 (2003). http://www.ncbi.nlm.nih.gov/pubmed/12862461. https://doi.org/10.1021/Ja034872n
43. Neese, F.: The ORCA program system. WIREs Comput. Mol. Sci. **2**(1) (2012). https://onlinelibrary.wiley.com/doi/abs/10.1002/wcms.81. Accessed 30 Sept 2021
44. Novoselov, K.S., Geim, A.K., Morozov, S.V., Jiang, D., Zhang, Y., Dubonos, S.V., Grigorieva, I.V., Firsov, A.A.: Electric field effect in atomically thin carbon films. Science (American Association for the Advancement of Science) **306**(5696), 666–669 (2004). https://www.science.org/lookup/doi/10.1126/science.1102896. Accessed 23 Sept 2021

45. Orlowski, G.A., Chowdhury, S., Long, Y.-T., Sutherland, T.C., Kraatz, H.-B.: Electrodeposition of ferrocenoyl peptide disul-fides. Chem. Commun. **2**(10), 1330–1332 (2005). http://www.ncbi.nlm.nih.gov/pubmed/15742068. https://doi.org/10.1039/B415278f
46. Paulus, G.L.C., Wang, Q.H., Strano, M.S.: Covalent electron transfer chemistry of graphene with diazonium salts. Acc. Chem. Res. **46**(1), 160–170 (2013). PMID: 22946516. https://doi.org/10.1021/ar300119z
47. Pawlowski, J., Juhaniewicz, J., Tymecka, D., Sek, S.: Electron transfer across α-helical peptide monolayers: importance of inter-chain coupling. Langmuir **28**(50), 17287–17294 (2012). http://www.ncbi.nlm.nih.gov/pubmed/23181704. https://doi.org/10.1021/La302716n
48. Perdew, J.P., Zunger, A.: Self-interaction correction to density-functional approximations for many-electron systems. Phys. Rev. B **23**(10), 5048–5079 (1981). https://link.aps.org/doi/10.1103/PhysRevB.23.5048. https://doi.org/10.1103/PhysRevB.23.5048
49. Perdew, J.P., Burke, K., Ernzerhof, M.: Generalized gradient approximation made simple. [Phys. Rev. Lett. **77**, 3865 (1996)]. Phys. Rev. Lett. **78**(7), 1396–1396 (1997). http://link.aps.org/doi/10.1103/PhysRevLett.78.1396
50. Perdew, J.P., Ruzsinszky, A., Csonka, G.I., Vydrov, O.A., Scuseria, G.E., Constantin, L.A., Zhou, X., Burke, K.: Restoring the density-gradient expansion for exchange in solids and surfaces. Phys. Rev. Lett. **100**(13), 136406 (2008). https://doi.org/10.1103/PhysRevLett.100.136406
51. Polo, F., Antonello, S., Formaggio, F., Toniolo, C., Maran, F.: Evidence against the hopping mechanism as an important electron transfer pathway for conformationally constrained oligopeptides. J. Am. Chem. Soc. **127**(2), 492–493 (2005). http://www.ncbi.nlm.nih.gov/pubmed/15643851. https://doi.org/10.1021/Ja043607e
52. Postma, H.W.: Rapid sequencing of individual DNA molecules in graphene nanogaps. Nano Lett. (American Chemical Society) **10**(2), 420–425 (2010). https://doi.org/10.1021/nl9029237. Accessed 11 Oct 2021
53. Prasongkit, J., Grigoriev, A., Pathak, B., Ahuja, R., Scheicher, R.H.: Transverse conductance of DNA nucleotides in a graphene nanogap from first principles. Nano Lett. (American Chemical Society) **11**(5), 1941–1945 (2011). https://doi.org/10.1021/nl200147x. Accessed 23 Sept 2021
54. Rocha, A.R., García-Suárez, V.M., Bailey, S., Lambert, C., Ferrer, J., Sanvito, S.: Spin and molecular electronics in atomically generated orbital landscapes. Phys. Rev. B **73**(8), 085414 (2006). https://link.aps.org/doi/10.1103/PhysRevB.73.085414
55. Shah, A., Adhikari, B., Martic, S., Munir, A., Shahzad, S., Ahmad, K., Kraatz, H.: Electron transfer in peptides. Chem. Soc. Rev. **44**, 1015–1027 (2015). https://doi.org/10.1039/C4CS00297K
56. Souza, F.A.L. de, Amorim, R.G., Scopel, W.L., Scheicher, R.H.: Electrical detection of nucleotides via nanopores in a hybrid graphene/h-BN sheet. Nanoscale (The Royal Society of Chemistry) **9**(6), 2207– 2212 (2017). https://pubs.rsc.org/en/content/articlelanding/2017/nr/c6nr07154f. Accessed 30 Sept 2021. https://doi.org/10.1039/C6NR07154F
57. Souza, F.A.L. de, Amorim, R.G., Scopel, W.L., Scheicher, R.H.: Controlled current confinement in interfaced 2D nanosensor for electrical identification of DNA. Phys. Chem. Chem. Phys. (The Royal Society of Chemistry) **21**(45), 24884–24890 (2019). https://pubs.rsc.org/en/content/articlelanding/2019/cp/c9cp03950c. Accessed 11 Oct 2021. https://doi.org/10.1039/C9CP03950C
58. de Souza, F.A.L., Amorim, R.G., Scopel, W.L., Scheicher, R.H.: Nano-structured interface of graphene and h-BN for sensing applications. Nanotechnology **27**(36), 365503 (2016). https://doi.org/10.1088/0957-4484/27/36/365503
59. Su, N.Q., Xu, X.: Development of new density functional approximations. Ann. Rev. Phys. Chem. **68**(1), 155–182 (2017). PMID: 28226221. https://doi.org/10.1146/annurev-physchem-052516-044835
60. Szabo, A., Szabo, J.: Modern Quantum Chemistry: Introduction to Advanced Electronic Structure Theory. Inglês. Revised ed. edição. Dover Publications, Mineola, N.Y. (1996)
61. Takeda, K., Morita, T., Kimura, S.: Effects of mono-layer structures on long-range electron transfer in helical peptide monolayer. J. Phys. Chem. B **112**(40), 12840–12850 (2008). http://www.ncbi.nlm.nih.gov/pubmed/18793017. https://doi.org/10.1021/Jp805711v

62. Vanderbilt, D.: Optimally smooth norm-conserving pseudopotentials. Phys. Rev. B **32**(12), 8412–8415 (1985). https://link.aps.org/doi/10.1103/PhysRevB.32.8412
63. Wang, Q.H., Shih, C.-J., Paulus, G.L.C., Strano, M.S.: Evolution of physical and electronic structures of bilayer graphene upon chemical functionalization. J. Am. Chem. Soc. **135**(50), 18866–18875 (2013). PMID: 24266808. https://doi.org/10.1021/ja4083914
64. Wang, Z.-Q., Lü, T.-Y., Wang, H.-Q., Feng, Y.P., Zheng, J.-C.: Review of borophene and its potential applications. Front. Phys. **14**(3), 33403 (2019). https://doi.org/10.1007/s11467-019-0884-5. Accessed 03 Oct 2021
65. Watanabe, J., Morita, T., Kimura, S.: Effects of dipole moment, linkers, and chromophores at side chains on long-range electron transfer through helical peptides. J. Phys. Chem. B **109**(30), 14416–14425 (2005). http://www.ncbi.nlm.nih.gov/pubmed/16852814. https://doi.org/10.1021/Jp051592g
66. Yu, M., Trinkle, D.R.: Accurate and efficient algorithm for bader charge integration. J. Chem. Phys. **134**(6), 064111 (2011)
67. Zhang, L., Yang, Z., Gong, T., Pan, R., Wang, H., Guo, Z., Zhang, H., Fu, X.: Recent advances in emerging Janus two-dimensional materials: from fundamental physics to device applications. J. Mater. Chem. A (The Royal Society of Chemistry) **8**(18), 8813–8830 (2020). https://pubs.rsc.org/en/content/articlelanding/2020/ta/d0ta01999b. Accessed 03 Oct 2021. https://doi.org/10.1039/D0TA01999B

Direct and Mediated Electron Transfer in Enzyme Electrodes

Marcus Victor Almeida Martins

1 Introduction

The first experimental reports on the origin of bioeletrochemistry were related to the work of Luigi Galvani, in October 1786. In his work, Galvani demonstrated that a frog's leg twitched upon contact with a metallic arch at the extremities of the leg. Named "Animal Electricity", this work is evidence of the origin of bioelectrochemistry long before electrochemistry itself [1–4]. Over 200 years after this experiment, bioelectrochemistry has advanced so as to be able to manipulate biomolecules, connect them with solid electrodes, and ultimately build and apply bioelectrochemical devices [5–8]. These biodevices' performance depends on understanding all physical and chemical properties of both the redox biomolecules and the electrodes that support enzyme immobilization. Thus, this chapter will discuss the processes of enzyme immobilization and electron transfer, focusing on the enzyme glucose oxidase.

2 Redox Enzyme—Glucose Oxidase

Several biological processes depend on electron transfer processes that occur between molecules undergoing oxidation and reduction. This redox process generates the energy needed for the development of biological functions or the construction of electrochemical biodevices [9]. In the latter case, several redox enzymes are extensively used on solid surfaces, generating an optimal interface for electron exchanges to occur [10]. Because they belong to the oxidoreductase class, these enzymes act as

M. V. A. Martins (✉)
Goiano Federal Institute – Catalão, Goiás, Brazil
e-mail: marccus.victor@ifgoiano.edu.br

F. N. Crespilho (ed.), *Advances in Bioelectrochemistry Volume 1*,
https://doi.org/10.1007/978-3-030-94988-4_2

Table 1 List of some redox enzymes and their active site

Enzyme	Active center	Ref.
Aldehyde dehydrogenase	Pyrroloquinoline quinone (PQQ)	[6]
Laccase oxidase	4 Cu (II) atoms	[11]
Alcohol dehydrogenase	Nicotinamide adenine dinucleotide (NAD)	[12]
Glucose dehydrogenase	Nicotinamide adenine dinucleotide (NAD)	[13]

excellent biocatalysts of specific substrates. Table 1 summarizes some redox enzymes and their active centers responsible for catalyzing their respective molecules.

One of the most classic enzymes used in bioelectrochemical devices is glucose oxidase (GOx), which often comes from *Aspergillus niger* fungi. GOx is a flavoprotein that catalyzes oxidation of β-D-glucose at its first hydroxyl group, using molecular oxygen as the electron acceptor, to produce D-glucono-delta-lactone and hydrogen peroxide. Sterically, GOx is dimeric, with its redox cofactor inside each monomer, called flavin adenine dinucleotid (FAD), as Fig. 1a, b show [14, 15].

Electrochemically, FAD undergoes oxidation according to the reaction in Fig. 2.

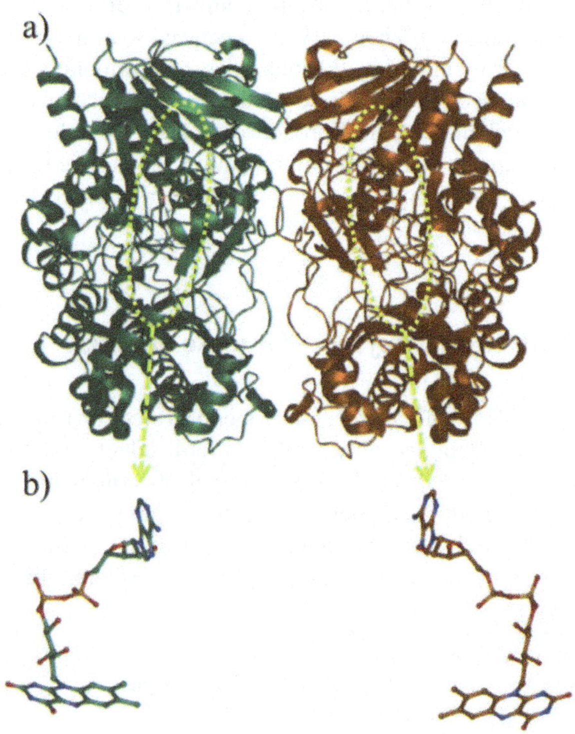

Fig. 1 **a** Computational representation performed on the protein data bank website for the GOx enzyme (1GPE) and for **b** FAD

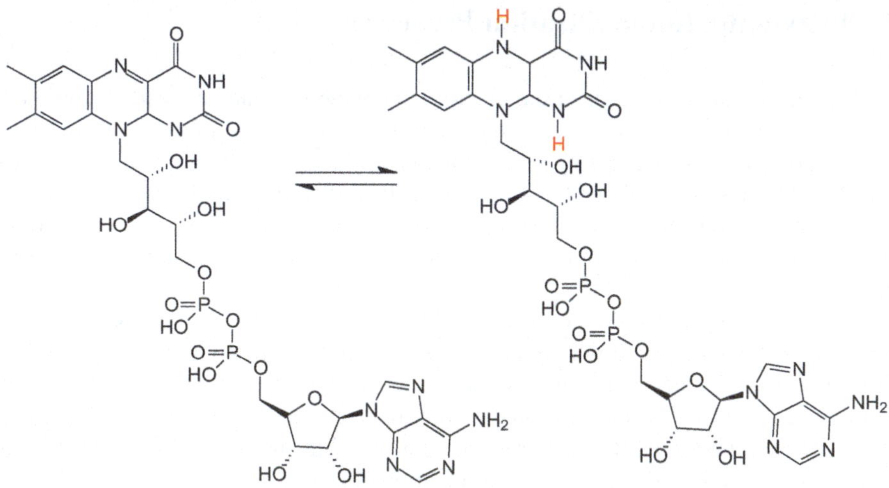

Fig. 2 Chemical structures of FAD and $FADH_2$

In the presence of two electrons and two hydrogen atoms, FAD, initially in the oxidized state, changes to the reduced state of $FADH_2$. This GOx redox cofactor reacts with glucose, according to the chemical reaction represented in Fig. 3 [15]. However, for an efficient electron transfer between the enzyme and its substrate on the solid surface of an electrode, the immobilization process becomes one of the biggest challenges for researchers, as we will discuss next.

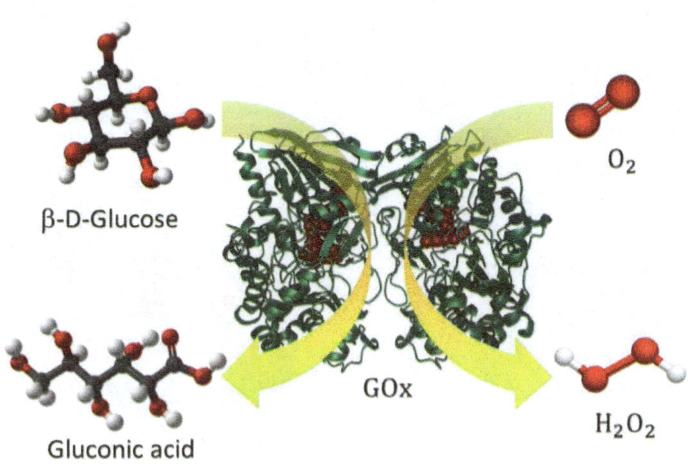

Fig. 3 **a** Schematic representation of the conversion of glucose into gluconic acid by the action of GOx-FAD

3 Enzymatic Immobilization Processes

The electron transfer process at the electrode/enzyme interface depends heavily on how the enzyme is immobilized on the electrode surface. The literature has classically reported five processes by which the enzyme can electronically connect with the electrode it is immobilized to. Each process has advantages and disadvantages for certain applications of this electrode biointerface. Figure 4 represents these immobilization processes: covalent bond, adsorption, cross-linking, electrostatic interaction, and affinity [16].

Immobilization through chemical bonding ensures strong fixation between the enzyme's functional groups and the electrode, thus promoting greater stability. However, the bioelectrode may generate a level of toxicity due to the chemical compounds used to fix the enzyme. In this respect, for in vivo applications, the presence of compounds that are bioincompatible with the environment can be an impediment to which the electrode will be applied [16].

Immobilization through physical adsorption can be considered the quickest and easiest of all methods. Another advantage is that chemical compounds are not needed to fix the enzyme, which makes the bioelectrode very applicable in in vivo implants. Nonetheless, in this technique the enzyme is weakly bound to the electrode surface, generating instability and inactivity of its active site [16].

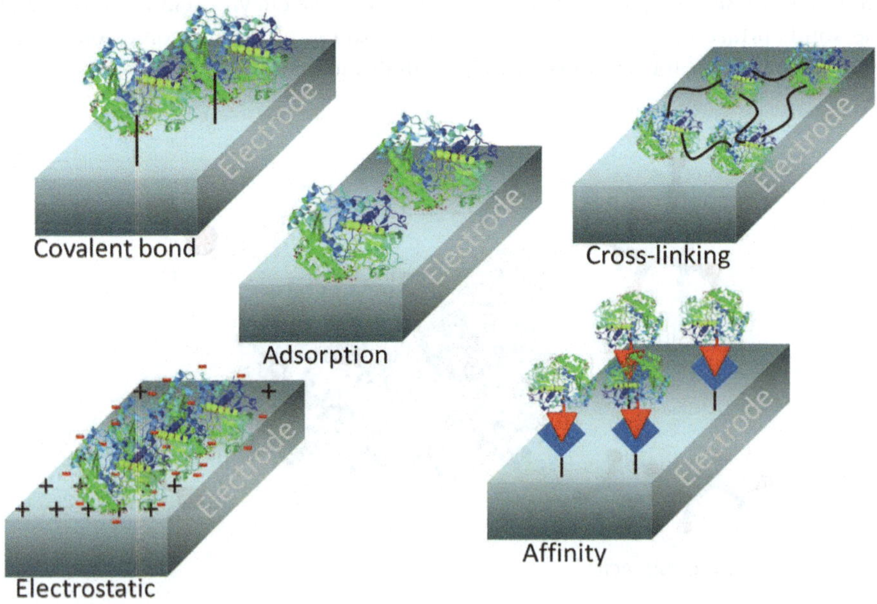

Fig. 4 **a** Schematic representation of the immobilization process in the electrode/enzyme interface

Both cross-linking and the biochemical affinity method require specific receptor substances and biomolecules, making the process a little more expensive and time-consuming. However, these processes guarantee electrode stability. In the case of affinity, other parameters must be considered, such as optimal pH of the medium and biocompatibility of the groups used to guarantee enzyme fixation on the electrode. Once immobilization on the electrode surface is completed, another aspect that must be investigated is the enzyme's structure and active site, as will be discussed below [16].

4 Direct Electron Transfer and Its Dependence on the Distance Between Enzyme and Electrode

A computer simulation performed on the PDB website with the structure of GOx (1GPE) showed that GOx exhibits its prosthetic group FAD in a non-centralized way inside each monomer, as Fig. 5 illustrates. The distances found from these simulations were 14.05, 19.36, 19.55, and 26.35 Å.

Some works have reported that the minimum optimal distance for the electron to tunnel from the enzyme to the electrode surface is 14 Å or less. In this sense, Fig. 7 shows four schematic representations (a–d) in which the enzyme succeeds or fails to exchange electrons with the electrode surface. As most immobilization processes do not guarantee the deposition of the enzyme through the face with the shortest distance, there is usually no direct electron communication at the electrode/enzyme interface. This is due to the volume of the enzyme's tertiary structure, which generates distances above 14 Å. This inhibits direct electron transfer, represented in Fig. 6a–c [17–19].

Martins and co-authors [19] presented an inherently effective strategy for direct electron transfer to occur using GOx and a graphene oxide-modified flexible carbon

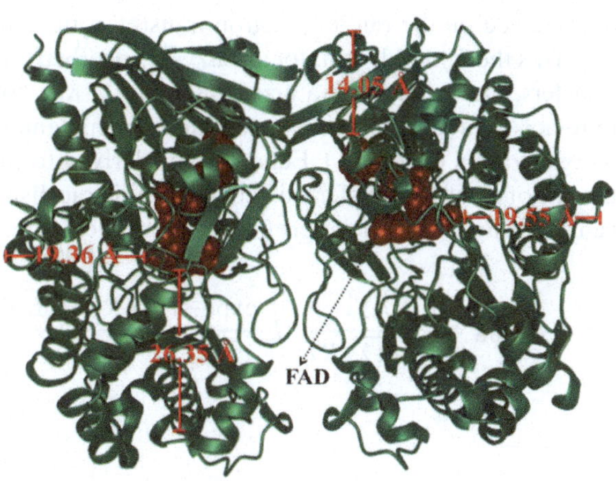

Fig. 5 **a** Schematic representation performed on the PDB website for GOx (1GPE)

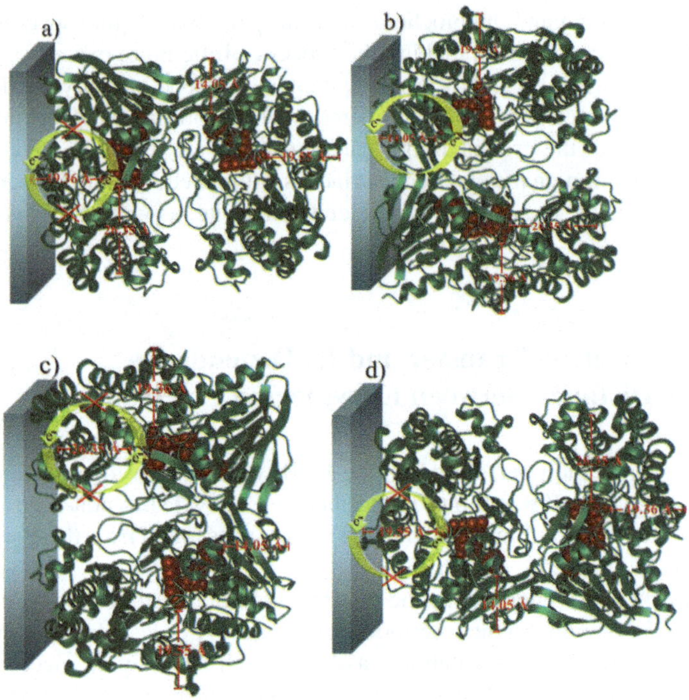

Fig. 6 Schematic representations **a–d** of the immobilization process through GOx's faces. Direct electron transfer only occurs when the enzyme adsorbs on the side that generates the shortest distance between FAD and the electrode surface

fiber (FCF) electrode. In their work, they showed that the presence of graphene oxide in the fiber structure shortens the distance between FAD and the electrode surface, ensuring fast and efficient electron exchange.

Marcus theory is another approach that can be considered in the study of distance dependence on the rate of electron transfer at the electrode/enzyme interface [20–22]. This theory holds that, for an enzyme adsorbed on a solid surface (homogeneous transfer), the oxidation k_{oxi} and reduction k_{red} rate constants require certain energy to re-accommodate all atoms under high overpotential conditions. This is known as reorganization energy (λ). Equation 1 is derived from Marcus theory and describes the rate of electron transfer using the Fermi–Dirac distribution, since electron transfer can occur to or from any Fermi level in the electrode:

$$k_{red/oxi} = \frac{k_{max}}{\sqrt{4\pi\lambda/RT}} \int_{-\infty}^{\infty} \frac{\exp(-1/4\lambda RT[\lambda \pm F\left(E - E^0\right) - RTx])}{1 + \exp(+)} dx \quad (1)$$

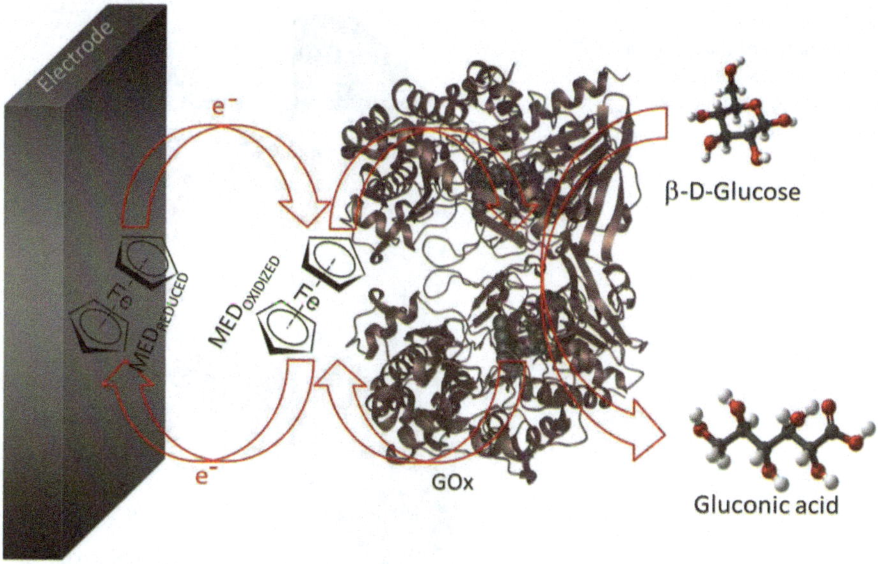

Fig. 7 Schematic representations of the electron transfer process between GOx enzyme and electrode surface mediated by ferrocene

where λ is the reorganization energy in eV (the energy required to reorient all atoms from the equilibrium state to the product state) and k_{max} is the maximum electron-transfer rate constant at high overpotential. Based on this theoretical simulation using the Marcus theory equation, the linear behavior obtained decreases linearly with increasing distance between the active enzymatic site and the electrode surface, as demonstrated by Marcus and co-authors [19].

5 Mediated Electron Transfer

Redox mediators are substances of low molar mass that can easily receive and donate electrons and thus facilitate the transfer of electrons between enzymes and the electrode surface. They are needed to mediate electrode/enzyme electron exchanges due to the distance between the redox site and the electrode. Some redox enzymes have a very bulky tertiary structure where the active site is located. This makes it impossible to tunnel the electron from the enzyme to the electrode. In this sense, numerous studies [18, 23–25] have reported the use of redox mediators in bioelectrodes. Depending on whether the application is in vivo or in vitro [7, 8], using mediator species in the electron exchange process may limit this application due to the toxicity that some mediators can generate. However, for biodevice applications that do not fall under these limitations, the great advantage of using a redox mediator is the ease of oxidizing and reducing the active site of an enzyme immobilized on an electrode

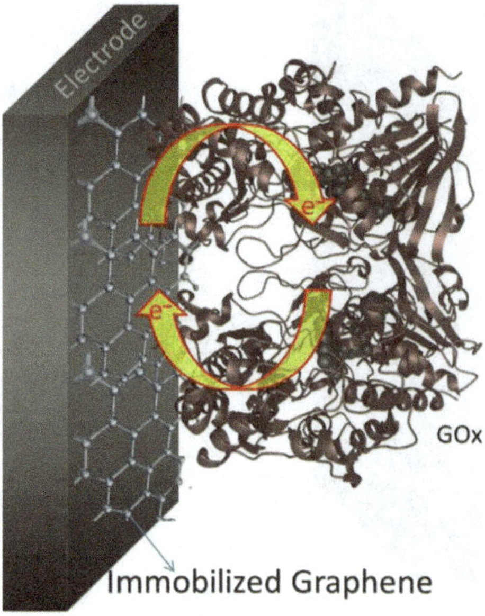

Fig. 8 Schematic representation of the electron transfer process between GOx enzyme and electrode surface mediated by graphene

surface. Figure 7 presents a scheme of how the redox mediator ferrocene acts in the electron exchange reaction between GOx and the electrode surface. Ferrocene both mediates electron exchange and enables FAD-GOx to react with more glucose molecules, generating gluconic acid as an end product.

Another advance in the field of mediated transfer at the electrode/enzyme interface is the use of nanomaterials (e.g., metallic nanoparticles [26], nanotubes [27], or graphene [19]). However, in this case, the communication between enzyme and electrode can be considered as facilitated rather than mediated. This consideration is because a nanomaterial generally does not have an oxidized and reduced state, as it occurs within the definition of a redox mediator species. Facilitation is due to the nanomaterial reducing the distance between the redox site and the electrode surface, as Fig. 8 illustrates.

Advances in the synthesis, development, manipulation, and functionalization of numerous advanced materials have also furthered the understanding of the properties of the electrode/enzyme interface applicable in bioelectrochemistry.

6 Conclusion

In this chapter, we discussed the physical and chemical properties at the electrode/enzyme interface, emphasizing GOx immobilized on a solid electrode surface. Some factors can limit charge transfer between an enzyme and an electrode. The distance between the enzymatic redox site and the electrode surface may be the main factor for mediated or direct transfer of electrons in a bioelectrode to occur. However, the presence of nanomaterials and/or redox mediating species at the electrode/enzyme interface can reduce this distance, facilitating electron communication between enzyme and electrode as well as the catalysis of specific substrates in the presence of these enzymes.

References

1. Piccolino, M.: Luigi Galvani and animal electricity: two centuries after the foundation of electrophysiology. Trends Neurosci. 443–448 (1997)
2. Galvani, L.: De viribus electricitatis in motu musculari commentarius. Bon. Sci. Art. Inst. Acad. Commun. 363–418 (1791)
3. Galvani, L.: Opere edite ed inedite del Professore Luigi Galvani raccolte e pubblicate dall'Accademia delle Science dell'Istituto di Bologna. Dall'Olmo, Bologna (1841)
4. Piccolino, M.: Animal electricity and the birth of electrophysiology: the legacy of Luigi Galvani. Brain Res. Bull. 381–407 (1998)
5. Gattani, A., Singh, S.V., Agrawal, A., Khan, M.H., Singh, P.: Recent progress in electrochemical biosensors as point of care diagnostics in livestock health. Anal. Biochem. 25–34 (2019)
6. Bollella, P., Gorton, L., Antiochia, R.: Direct electron transfer of dehydrogenases for development of 3rd generation biosensors and enzymatic fuel cells. Sensors **18**, 1319 (2018)
7. Gross, A.J., Holzinger, M., Cosnier, S.: Buckypaper bioelectrodes: emerging materials for implantable and wearable biofuel cells. Energy Environ. Sci. **11**, 1670–1687 (2018)
8. Karim, N.A., Yang, H.: Mini-review: recent technologies of electrode and system in the enzymatic biofuel cell (EBFC). Appl. Sci. **11**, 5197 (2021)
9. Zhou, M.: Recent progress on the development of biofuel cells for self-powered electrochemical biosensing and logic biosensing: a review. Electroanalysis **27**, 1786–1810 (2015)
10. Léger, C., Bertrand, P.: Direct electrochemistry of redox enzymes as a tool for mechanistic studies. Chem. Rev. **108**, 2379–2438 (2008)
11. Mate, D.M., Alcalde, M.: Laccase: a multi-purpose biocatalyst at the forefront of biotechnology. Microb. Biotechol **10**, 1457–1467 (2016)
12. Franco, J.H., Minteer, S.D., Andrade, A.R.: Product analysis of operating an ethanol/O_2 biofuel cell shows the synergy between enzymes within an enzymatic cascade. J. Electrochem. Soc. **165**(9), 575–579 (2018)
13. Stolarczyk, K., Rogalski, J., Bilewicz, R.: NAD(P)-dependent glucose dehydrogenase: applications for biosensors, bioelectrodes, and biofuel cells. Bioelectrochemistry **135**, 107574 (2020)
14. Ferri, S., Kojima, K., Sode, K.J.: Review of glucose oxidases and glucose dehydrogenases: a bird's eye view of glucose sensing enzymes. Diabetes Sci. Technol. **5**, 1068–1076 (2011)
15. Bankar, S.B., Bule, M.V., Singhal, R.S., Ananthanarayan, L.: Glucose oxidase—an overview. Biotechnol. Adv. **27**, 489–501 (2009)
16. Sassolas, A., Blum, L.J., Leca-Bouvier, B.D.: Immobilization strategies to develop enzymatic biosensors. Biotechnol. Adv. **30**, 489–511 (2012)
17. Milton, R.D., Minteer, S.D.: Direct enzymatic bioelectrocatalysis: differentiating between myth and reality. J. R. Soc. Interface **14**, 20170253 (2017)
18. Utterback, J.K., Ruzicka, J.L., Keller, H.R., Pellows, L.M., Dukovic, G.: Electron transfer from semiconductor nanocrystals to redox enzymes. Annu. Rev. Phys. Chem. **71**, 335–359 (2020)
19. Martins, M.V.A., Pereira, A.R., Luz, R.A.S., Iost, R.M., Crespilho, F.N.: Evidence of short-range electron transfer of a redox enzyme on graphene oxide electrodes. Phys. Chem. Chem. Phys. **16**, 17426–17436 (2014)

20. Marcus, R.A.J.: On the theory of oxidation-reduction reactions involving electron transfer. Chem. Phys. **24**, 966–978 (1956)
21. Marcus, R.A.: Chemical and electrochemical electron-transfer theory. Ann. Rev. Phys. Chem. **15**, 155–196 (1964)
22. Marcus, R.A., Sutin, N.: Electron transfers in chemistry and biology. Biochim. Biophys. Acta. **811**, 265–322 (1985)
23. Hitaishi, V.P., Clement, R., Bourassin, N., Baaden, M., Poulpiquet, A., Sacquin-Mora, S., Ciaccafava, A., Lojou, E.: Controlling redox enzyme orientation at planar electrodes. Catalysts **8**, 192 (2018)
24. Xiao, X., Yan, X., Magner, E., Ulstrup, J.: Polymer coating for improved redox-polymer-mediated enzyme electrodes: a mini-review. Electrochem. Commun. **124** 106931 (2021)
25. Yuan, M., Minteer, S.D.: Redox polymers in electrochemical systems: from methods of mediation to energy storage. Curr. Opin. Electrochem. **15**, 1–6 (2019)
26. Paul, A., Vyas, G., Paul, P., Srivastava, D.N.: Gold-nanoparticle-encapsulated ZIF-8 for a mediator-free enzymatic glucose sensor by amperometry. ACS Appl. Nano Mater. **1**, 3600–3607 (2018)
27. Xia, H., Zeng, J.: Rational surface modification of carbon nanomaterials for improved direct electron transfer-type bioelectrocatalysis of redox enzymes. Catalysts **10**, 1447 (2020)

A Case Study of Cytochrome c

Eliemy F. S. Bezerra, Caio Lenon C. Carvalho, Everson T. S. Gerôncio, Welter Cantanhêde, and Roberto A. S. Luz

1 Electron Transfer Reactions

Electron transfer (ET) reactions are key steps in many biological processes, such as photosynthesis, metabolism and cell respiration [1, 2]. Furthermore, the deeper understanding of such electron transfer processes is a fundamental issue in biotechnological applications, included enzymatic fuel cells, and third-generation biosensors [3–5]. ET reactions between proteins and electrode surfaces typically occur in two ways. The first is based on the use of redox mediators, which promote a mediated electron transfer (Fig. 1a). In this case, the mediator must be able to transfer electrons quickly with the application of a small driving force and commonly its reduction potential is what dictates the operating potential of electrode. The second way is based on direct electron transfer (DET) between the protein active center and the electrode surface, without the need of redox mediators or co-substrates (Fig. 1b). This DET may be established by using an electrode optimized configuration, where the protein must be strongly adsorbed with its active site facing the electrode surface, in order to ensure the shortest possible ET distance, generally not exceeding 14 Å as demonstrated by Dutton and coworkers [6]. However, in many cases this distance is increased due to the bulky tertiary or quaternary structure of proteins preventing DET. In addition, after immobilization, the protein chain denaturation with subsequent electrode surface passivation is often observed. In this context, a third way that has been extensively explored in recent years to maximize ET is the use of nanomaterials at the protein/electrode interface [7–9] (Fig. 1c). In this case, although the electron transfer occurs through the nanomaterial located at the enzyme/electrode

E. F. S. Bezerra · E. T. S. Gerôncio · R. A. S. Luz
PPGQ-GERATEC, State University of Piauí, Teresina, PI 64002-150, Brazil

C. L. C. Carvalho · E. T. S. Gerôncio · W. Cantanhêde · R. A. S. Luz (✉)
Department of Chemistry, Federal University of Piauí, Teresina, PI 64049-550, Brazil
e-mail: robertoluz@ufpi.edu.br

F. N. Crespilho (ed.), *Advances in Bioelectrochemistry Volume 1*,
https://doi.org/10.1007/978-3-030-94988-4_3

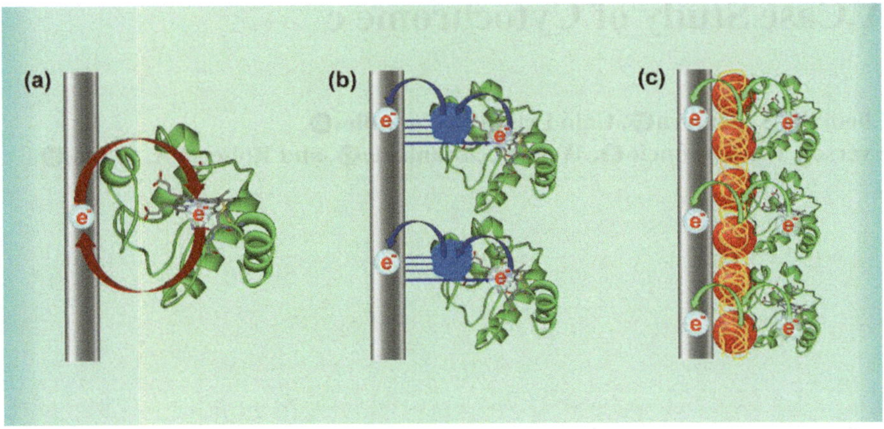

Fig. 1 Schematic representation of electron transfer electrochemical process on redox proteins modified electrodes. **a** DET between enzyme and electrode surface; **b** Mediated electron transfer; **c** Electron transfer between enzyme and the surface of electrodes modified with nanomaterials

interface, many authors have considered this to be DET [10–13], once the immobilized nanomaterials usually act as an extension of the electrode surface. In addition, this strategy promotes an increase in surface area to be covered and can lead to enhanced stability of the enzymatic electrode.

Direct electrochemistry of immobilized proteins is quite advantageous over solution studies, since diffusion processes are avoided and the electron transfer rate constant can be constrained as a function of the distance from redox site to electrode surface and of reaction driving force. Electrochemical techniques, such as cyclic voltammetry, combined with choice and proper treatment of electrodes, have allowed the obtaining of thermodynamic and kinetic variables related to heterogeneous electron transfer processes of adsorbed molecules [14, 15].

The pioneering works on DET of adsorbed redox proteins were published independently by Eddowes and Hill [16] and Yeh and Kuwana [17] in 1977. In these studies, the authors investigated, by cyclic voltammetry, the redox properties of cytochrome c (Cyt c) on gold and tin-doped indium oxide (ITO) electrodes, respectively. These pioneering works demonstrated that important information about redox properties of proteins could be obtained from experiments where the DET is established. Since then, several researchers have developed strategies to promote a strong adsorption of proteins on solid electrodes, in order to preserve the proteins native properties and provide a suitable environment for investigating direct electron transfer. In this sense, Armstrong and coworkers [14, 15, 18] have investigated by voltammetry, the charge transfer reactions, elucidating several redox protein catalytic mechanisms, paving the way for a new approach called "protein film voltammetry" (PFV).

Among various redox-active proteins, Cyt c stands out as one of the most used in ET studies, due to its unique characteristics and electrochemical properties, as discussed in the following topic.

2 Cytochrome c, a Brief Understanding

The Cyt c is a cationic highly water-soluble protein, with approximately spherical shape (relatively diameter small of 3.4 nm), containing a prosthetic group formed by the porphyrin complex of heme (Fe^{2+}) and amino acid residues (often cysteine, histidine and methionine) as can be seen in Fig. 2 [19–22].

The redox cofactors (heme centers Fe) play essential role as efficient electron carriers from complex III (quinol: cytochrome *c* oxidoreductase or cytochrome *bc1* complex) for complex IV (cytochrome *c* oxidase) in the inner mitochondrial membrane [21, 22]. Thus, the Cyt c is able to reversibly alternate between Cyt c Fe^{3+} and Cyt c Fe^{2+} states during ET on the electrode surface. The respiratory complexes catalyze electron transfer reactions by reduction or oxidation selective of substrates, with translocation of protons from the negative to the positive side of the membrane (Fig. 3a). In particular, the end of the aerobic respiratory chain the cytochrome c oxidase catalyze the reduction of oxygen (final electron acceptor) to water [23]. Figure 3b shows a proposed mechanism of ET from cytochrome c into direct electrical contact with the electrode surface [21–23].

An interesting feature observed for Cyt c is distance of ~17 Å for an optimum orientation of the redox active site (parallel orientation of heme plane) in relation to electrode surface, this value is close to the ideal distance (14 Å) considered for an efficient DET to occur [22, 25]. Furthermore, the Cyt c exhibit low cost, high stability, electrochemical activity over a wide pH range (pH 2–11) and, unlike many other redox proteins, its active center is not buried as deeply in the amino acid sequence, making it an ideal protein for DET studies [26, 27].

Despite the favorable characteristics of Cyt c to electron transfer, DET is not so simple to be achieved, as this process is directly related to the way in which the enzyme is immobilized on electrode surface. In this context, different enzymatic

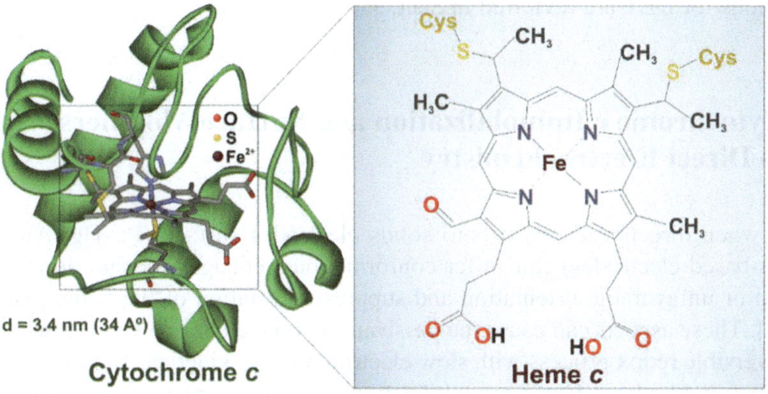

Fig. 2 Three-dimensional (3D) structure of cytochrome c and tetrapyrrolic macrocycle ligated to an iron cation (heme c prosthetic group). Adapted and reprinted with permission from Refs. [20, 22]. Copyright 2021 Elsevier

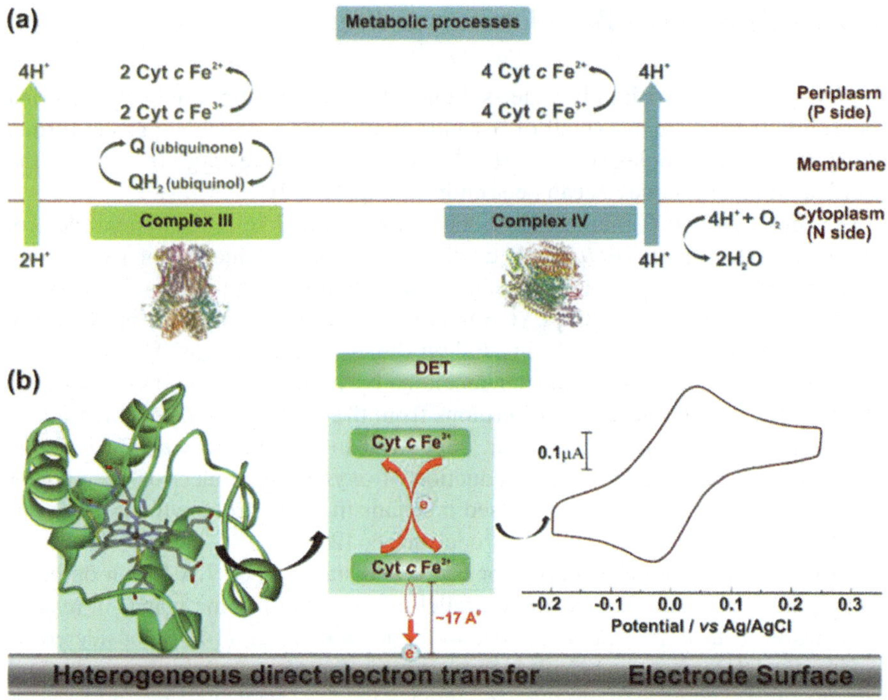

Fig. 3 **a** Schematic representation of electron transport in the active sites of respiratory complexes III and IV [1–3]. **b** Schematic illustration of the DET process in electrode surface and typical cyclic voltammetric of the cytochrome c at the formal potential of 9 mV versus Ag/AgCl. Adapted and reprinted with permission from Ref. [24]. Copyright 2021 Royal Society of Chemistry

immobilization strategies and electrode surface modifications to achieve favorable conditions for DET are reviewed in Sect. 3.

3 Cytochrome c Immobilization and Surface Modifiers to Direct Electrochemistry

Cyt c when directly adsorbed onto solids electrodes surface (Pt, Hg, Au, Ag or carbon-based electrodes) can suffer conformational changes (denaturation of their chains) or unfavorable orientation and suppressed rotation of the heme plane [10, 19, 26]. These aspects can cause the passivation of the electrode surface, which lead to irreversible redox process with slow electron transfer kinetics. To minimize these problems, methods of Cyt c immobilization onto conductive support are strategically selected for promote DET. In this case, the Cyt c immobilization may keep their functional and structural properties and improve the electrical communication between the Cyt c Fe^{2+}/Cyt c Fe^{3+} center and the electrode surface. To date, the main methods of Cyt c immobilization are physical adsorption, encapsulation/entrapment, covalent anchorage and layer-by-layer (LbL) assembly, as show in Fig. 4 [4, 10, 23, 26]. The structural characteristics of Cyt c play an important role in its immobilization mechanism onto electrodes. Thus, at neutral pH, lysine and arginine residues contribute to a net positive charge, which makes electrostatic interaction with the negatively charged electrode surface possible [26].

The selection of immobilization method depends on its advantages, disadvantages and type of interaction, as they affect the physicochemical properties of the Cyt c/electrode interface. Furthermore, most methods involve essential parameters for efficient application, for example ease of operation, storage stability, sensitivity, selectivity, reproducibility and stability for long-term performance [28]. For these reasons, the choice of the ideal method is a very complex issue and, therefore, in some cases the strategy selection is result of a combination of one or more immobilization types [23, 28].

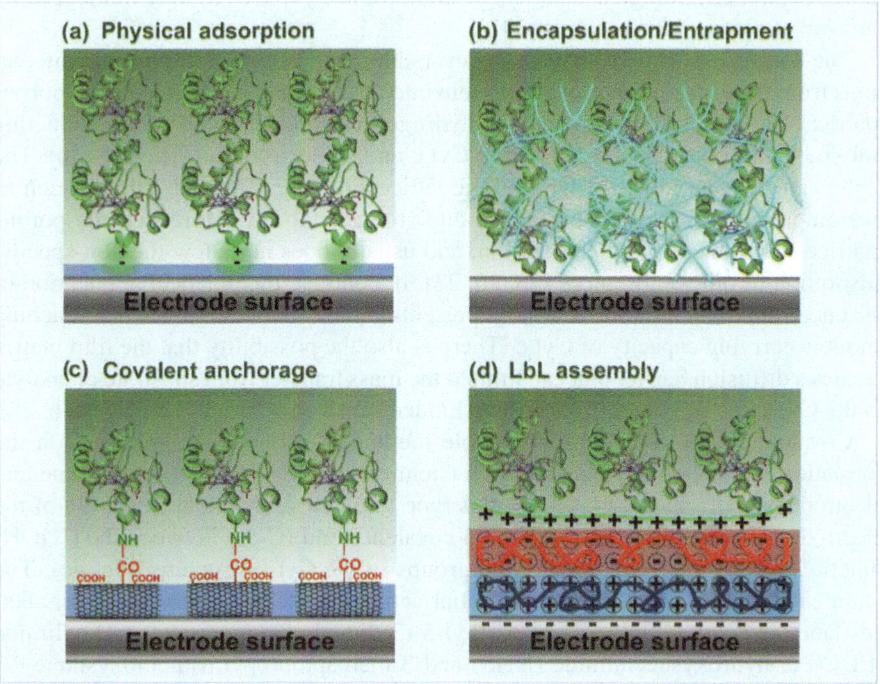

Fig. 4 Schematic illustration of Cyt c immobilization methods onto electrodes surface. **a** physical adsorption based on the electrostatic interaction between the negatively charged electrode and positively charged groups of Cyt c; **b** encapsulation/entrapment into hydrogel or polymers porous matrices; **c** covalent binding (C–N) onto COOH-functionalized single-walled carbon nanotubes deposited on electrode surface; **d** LbL assembly with negatively charged substrate for sequential deposition of polyanions, polycations and Cyt c via electrostatic interaction to formed a multilayer structure. Adapted and reprinted with permission from Ref. [26]. Copyright 2021 Elsevier

The physical adsorption is one method of reversible immobilization based on non-covalent interactions (non-specific forces), included π–π interactions, van der Walls forces, hydrophobic interactions, hydrogen bonds and ionic bonding (salt linkages) [26, 28]. Adsorption also can occur through electrostatic interaction between positive residues of the Cyt c and the negatively electrode surface (Fig. 4a). The main advantages of the physical adsorption are simplicity and operational speed (few steps), low-cost and the Cyt c active site is normally unaffected [28]. Nonetheless, the main disadvantage is the low operational and storage stability due to desorption of Cyt c from the electrode under changing conditions of temperature, pH and ionic strength. Another disadvantages include denaturation of the Cyt c (resultant of direct contact between Cyt c/electrode) and non-specific adsorption of other proteins or substance [26, 28]. Recently, it has been reported that the physical adsorption by hydrophobic interaction affects the reactivity of immobilized Cyt c [20]. In this study, species derived from yeast Cyt *c* (ycc) were immobilized onto a gold electrode surface coated with a hydrophobic layer of decane-1-thiol. The structural and functional properties (catalytic and physiological activity) of immobilized ycc are strongly influenced by the type of physical interactions [29].

The entrapment is considered an irreversible method based on physical interactions from the encapsulation of the protein into three-dimensional network or porous matrices [26, 28]. In this case, gelatin hydrogels or conducting and non-conducting polymers matrix can be used to entrap Cyt c on the electrode surface (Fig. 4b). The Cyt c confinement inside of hydrophilic lattice structure offers advantages such as biocompatibility, improve operational and storage stability. Moreover, the porous matrices minimize Cyt c denaturation, and usually does not allow the non-specific adsorption of other substances [23, 26, 28]. In contrast, the practical application of the encapsulation/entrapment method presents some disadvantages such as leaching and low carrying capacity of Cyt c. There is also the possibility that the film matrix creates a diffusion barrier that can limited the mass transfer from substrate or analyte to the Cyt c Fe^{2+}/Cyt c Fe^{3+} center and hinder the electrochemical process [26, 28].

Covalent anchorage is an irreversible method of immobilization based on the formation of stable complexes through chemical binding between the enzyme and electrode support matrix [4, 23]. In this type of method, chemical activation of the electrode surface occurs first to form a covalent bond (C–N) between the COOH-functionalized support and the amine groups in the Cyt c structure (Fig. 4c). The main chemical coupling agents for initial activation include glutaraldehyde, alkanesilanes and carbodiimides, e.g., 1-ethyl-3-(3-dimethylaminopropyl) carbodiimide (EDC), N-hydroxysuccinimide (NHS) and 3-mercaptopropyl trimethoxysilane (3-MPTMS). These chemically modified surfaces can undergo chemical coupling with functional groups from the Cyt c amino acid residues, such as side chains containing COOH (aspartic and glutamic), NH_2 (lysine) or SH (cysteine) groups [23, 26, 28]. The chemical interaction between electrode surface electrophilic groups and nucleophiles on Cyt c provides some advantages, for example, high bond stability, absence of diffusion barriers and short response times [4, 23]. However, covalent immobilization often results in some disadvantages, including changes in protein conformation, loss of biocatalytic activity, poor electrode reproducibility, and use of toxic coupling

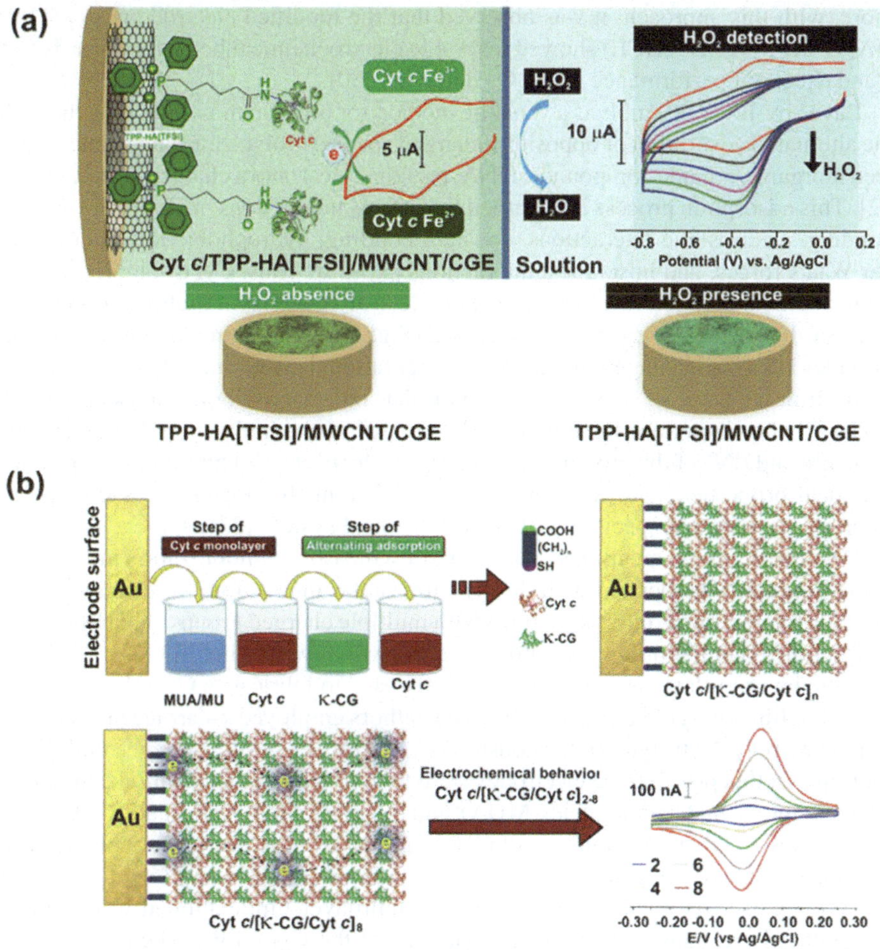

Fig. 5 **a** Schematic illustration of cytochrome *c* immobilized in TPP-HA[TFSI]/MWCNT/GCE, and cyclic voltammograms in the presence of H_2O_2. **b** Scheme of steps of assembly of κ-CG and Cyt *c* for form multilayer structure ([κ-CG/Cyt *c*]$_n$) on a Au electrode surface modified with a Cyt *c* monolayer, and voltammetric response of the multilayers electroactive Cyt *c*, [κ-CG/Cyt *c*]$_n$ (n = 2, 4, 6, 8). Adapted and reprinted with permission from Refs. [30, 36]. Copyright 2021 Elsevier

agents [23, 28]. Murphy and coworkers reported an approach to covalent immobilization of Cyt c from COOH-functionalized ionic liquid (TPP-HA [TFSI]) and multiwalled carbon nanotube (MWCNT) [30]. In this work, TPP-HA [TFSI] was deposited onto MWCNT-modified glassy carbon electrode (GCE) surface, followed by chemical anchorage of Cyt c using EDC, coupled to NHS (EDC/NHS). Thus, EDC/NHS was applied as the coupling agent to form stable amide binding (C–N) between COOH groups of TPP-HA[TFSI] and NH_2 groups of Cyt c (Fig. 5a). Furthermore, with this approach, it was observed that the modified electrode (Cyt *c*/TPP-HA[TFSI]/MWCNT/GCE) showed favorable electrochemical behavior towards Cyt *c* and excellent performance for H_2O_2 detection [30].

Layer-by-layer assembly is a versatile method for protein immobilization through the alternated adsorption of oppositely charged species (polyelectrolytes, functionalized inorganic/organic compounds, DNA, proteins, etc.) onto a charged substrate [31, 32]. This adsorption process is governed by various interactions, including hydrogen bonding, electrostatic interactions, covalent bonding, hydrophobic interaction, van der Waals forces, and host–guest interactions [31–33]. Thus, Cyt c soluble in water with chemical charged residues act as polyelectrolytes to form multilayer architectures in the surface of a charged electrode (Fig. 4d). This immobilization method provides many advantages, including: (I) operational ease, and (II) compatibility with different types of substrates (flat/non-flat and macroscopic/nanoscopic electrodes). Other important features are (III) control of the multilayers homogeneous structure and (IV) of the proteins amount on the electrode, (V) fine-tuning of physic-chemical properties, and (VI) deposition of different (bio)materials as strategy to improve stability and specificity of electrochemical system [33, 34].

Despite the above-mentioned characteristics, the LbL immobilization method has some limitations associated with the type molecular interaction, for example, electrostatic interaction requires structures with multiple charged groups, and H-bonding interaction generally produces multilayer films with low stability [33–35]. The electrostatic layer-by-layer assembly was recently used to fabricate a Cyt c-based multilayer thin film on a gold electrode [36]. The authors employed κ-carrageenan (κ-CG) as anionic polyelectrolyte for the construction of fully electroactive Cyt c multilayer biofilms. In this process, the electrodes were firstly incubated in solution containing 11-mercapto-1-undecanoic acid (MUA) and 11-mercapto-1-undecanol (MU), and then it were incubated in solution of Cyt c to form a redox protein monolayer. The multilayer κ-CG/Cyt c structure was fabricate by alternating adsorption of κ-CG and Cyt c (Fig. 5b). As results, the κ-CG/Cyt *c* multilayer film exhibited well-defined voltammetric response related to ET reactions of the Cyt c [36]. The combination of SAMs with layer-by-layer assembly modified electrochemical performance of the as-prepared multilayers thin films. This was observed by different amounts of immobilized electroactive Cyt *c* on the electrode surface (τ, pmol cm^{-1}) and for the electron transfer rate constant (k_{ET}, s^{-1}). Thus, the κ-CG-based Cyt c monolayer film and 8-bilayer κ-CG/Cyt c structure (Cyt c/[κ-CG/Cyt c]$_8$) showed the τ values of 10.87 and 158.01 pmol cm^{-1}, respectively. These results indicates that κ-CG/Cyt c multilayer film exhibited the concentration of electroactive Cyt c species ~15 folds greater than that on electrode modified with Cyt c monolayer [36].

4 Direct Electrochemistry of Cyt c at Nanomaterials-Modified Electrodes

Despite the advantages of immobilization methods, in some cases, the adsorbed Cyt c presents difficulties in reaching DET on a conductive surface. Unfavorable orientation on unmodified electrode, low surface coverage of electroactive species, and critical electron transfer distance between the electrode surface and the redox center are the main factors limiting the electrochemical processes of Cyt c [37]. Nowadays, the electrode modification with nanomaterials is one of the main strategies to overcome these limitations [38]. This approach allows electronically connecting the redox centers of the protein with the electrode surface. Thus, the nanomaterial acts as an electrical "wiring" (nano-bridges) to promote an efficient electronic communication. Moreover, the ultra-small size and structural properties of nanomaterials may also allow the creation of a favorable microenvironment for Cyt c (in terms of stability and increased protein load) improving the direct electron transfer [38]. The main classes of these electrode modifiers include metallic nanoparticles, carbon-based nanomaterials, mesoporous nanostructures, and nanocomposites [7–9]. The type of nanostructure formed influences the physical and chemical properties of the electrode surface, as well as the nature of the protein/nanomaterial interface interactions. In this section, some characteristics for DET of Cyt c on nanomaterials-modified electrodes are described.

Metal-based nanomaterials show interesting advantages such as high electrical conductivity, large surface area, high potential, electronic and electrocatalytic properties [39]. The use of metallic nanoparticles for electrode modification can improve the Cyt c electron transfer process. This improvement in electrochemical behavior can be observed by increasing the electron rate constants, increasing the surface coverage and decreasing the distance between iron center and electrode surface [10, 26, 39]. Kosopova et al. investigated the effect of different types of gold nanostructures on the Cyt *c* direct electrochemical [40]. Nanostructures of oleylamine-stabilized monodisperse spherical nanoparticles (OA/AuNPs) and ultrathin gold nanowires (OA/AuNWs) were assembled on gold electrode surfaces to immobilize Cyt c. As results, the thin-film bioelectroactive nanoarchitectures showed electrochemical properties depending on nanostructure type and the method of assembly onto electrode surface. Considering the same assembly strategy, the effect of nanoparticles geometry was observed by change of ca. −20 mV at the redox potential (Cyt c Fe^{2+}/Cyt c Fe^{3+}) of the OA/AuNWs/electrode compared to OA/AuNPs electrode. Furthermore, OA/AuNWs/Cyt *c* bioelectrode improves the electron transfer kinetics when compared to OA/AuNPs/Cyt c system, with k_{ET} values of 2.4 ± 0.5 and 2.0 ± 0.3 s^{-1}, respectively [40]. In other work, DET of Cyt *c* was studied in a glassy carbon electrode (CGE) modified by 3D graphene aerogel (3DGA) decorated with gold nanoparticles (AuNPs) [41]. The 3D-networked structure (3DGA-AuNPs) with large specific surface area and high conductivity facilitated the DET between iron center and electrode (Fig. 6a). The nanostructured electrode (3DGA-AuNPs/Cyt *c*/GCE) showed a pair of well-defined and quasi-reversible redox peaks for Cyt *c*

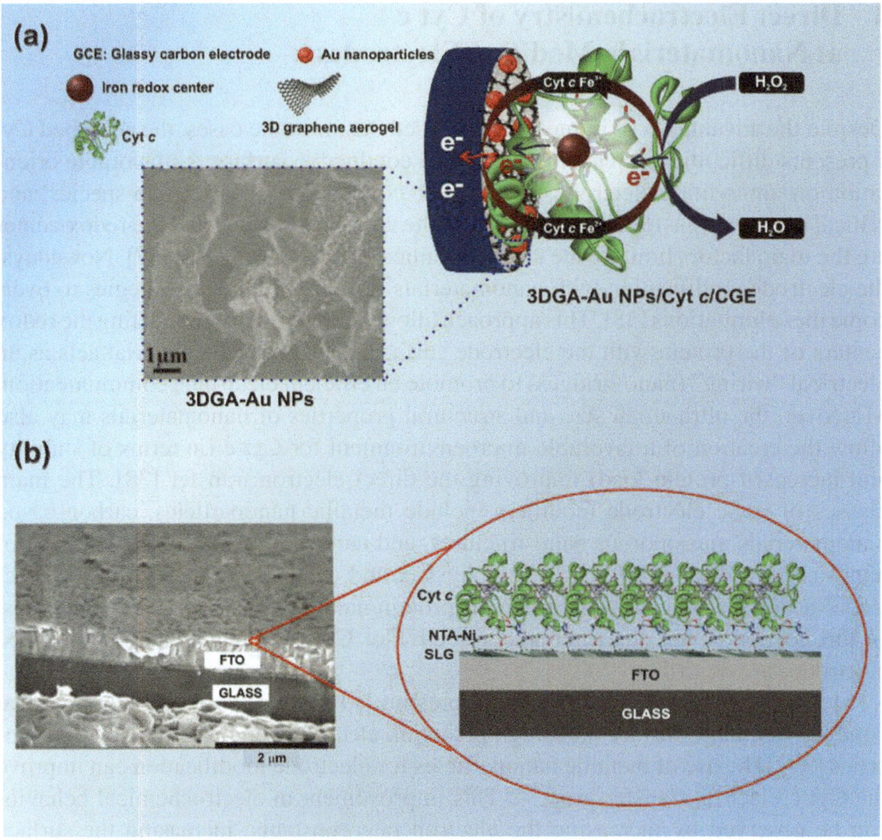

Fig. 6 **a** Field-emission scanning electron microscope image of 3DGA-Au NPs and schematic representation of electrocatalytic mechanism for determination of H_2O_2 in the 3DGA-Au NPs/Cyt *c*/CGE surface. **b** Cross-sectional scanning electron microscopy image of glass/FTO substrate covered with SLG/pyr-NTANi/cyt c_{553} and illustration of FTO/SLG/pyr-NTANi/ cyt c_{553} nanostructure. Adapted and reprinted with permission from Refs. [41, 45]. Copyright 2021 Elsevier

Fe^{2+}/Cyt c Fe^{3+} process and efficient electrocatalytic behavior toward H_2O_2 sensing [41].

Carbon-based nanomaterials (carbon nanotubes, graphene and its derivatives) are one of the most widely used nanoscale materials to form an electrochemical interface that improves the protein electron transfer [42]. This fact is due to its desirable physicochemical properties, such as improved electrical conductivity, extraordinary electrical character, porous structure, mechanical strength, chemical stability and large specific surface area [7, 32, 43]. Multi-walled carbon nanotube-modified electrodes can improve the electrochemical behavior for Cyt c DET [44]. In this case, the presence of carbon nanotubes promotes conformational changes in the protein at various nano-levels, including secondary structure, spatial orientation, and spin states. The change in secondary structure and transition of spin states provide an optimum orientation of the redox center and relatively high energy to create a microenvironment required to facilitate DET of Cyt c [44].

Very recently, Niewiadomski and coworkers investigated the direct electron transfer of the novel cytochrome c553 peptide linker variants (Cyt c553) immobilized on the graphene-modified electrode [45]. The fluorine doped tin oxide (FTO) electrode was covered with a single layer graphene (SLG) functionalized with a pyrene derivative self-assembled monolayer coordinated with Ni^{2+} species (FTO/SLG/pyr-NTANi) [45]. The conductive bioelectrode (FTO/SLG/pyr-NTANi/Cyt c_{553}) showed a well-defined nanostructure with formation of a stable Cyt c_{553} monolayer (Fig. 6b). These studies have demonstrated that presence of C-terminal His6-tag of Cyt c_{553} were crucial for optimized heme orientation on FTO/SLG/pyr-NTANi interface. Moreover, the fine-tuning of the prepared nanoarchitectures electronic properties was influenced by optimization of the heme group inclination angle and by the optimal HOMO/LUMO levels of the iron center interaction at the interface. These changes favor an efficient DET of cyt c553 with a 20-fold increase in cathodic photocurrent when compared to similar electrodes previously reported [45].

Mesoporous nanostructures (e.g., ZnO, TiO_2, SnO_2, Co_3O_4, Fe_3O_4, SiO_2, Nb_2O_5) are of great importance as electrode modifiers to investigate DET of proteins. Due to its various properties such as high surface-to-volume ratio, ultra-high surface areas, large pore volumes, unique structural stability, electrical activity, good biocompatibility, catalytic properties, adjustable pore sizes and shapes [7, 46]. The combination of these advantages creates a suitable microenvironment to facilitate Cyt c/electrode interactions and promote DET on electrode surface. A mesoporous structure constituted by cobalt metal–organic frameworks/gold nanoparticles (Co-MOFs/AuNPs) was used for immobilize a nanosystem based on cytochrome *c*-multiwalled carbon nanotubes (Cyt *c*-MWCNTs) [47]. In this work, Huang and coworkers studied the electrocatalytic activity of a novel electrochemical nanoplatform (Cyt c-MWCNTs/AuNPs/Co-MOFs/CGE) for nitrite detection. The Co-MOFs/AuNPs showed a lamellar structure with large specific surface area for immobilize and improve the electron transfer of Cyt *c*-MWCNTs. As results, Cyt c-MWCNTs/AuNPs/Co-MOFs/CGE electrode exhibited a quasi-reversible redox pair for Cyt c Fe^{2+}/Cyt c Fe^{3+} process and improved electrocatalytic activity towards nitrite (with low detection limit, high selectivity and good reproducibility) [47].

Melo et al. reported the preparation of magnetically induced mesoporous framework (Fe_3O_4–AuNPs) as support for Cyt *c* immobilization [48]. The authors demonstrated thin film formation of Fe_3O_4–AuNPs derived from the magnetic assembly of magnetite nanoparticles decorated with nearly monodisperse gold nanoparticles (functionalized with L-cysteine (cyst) molecules) on an ITO electrode surface. This (ITO/Fe_3O_4–AuNPs/Cyst) film exhibited a thickness of approximately 5-μm, electrochemically active surface area (EASA) of 71.87 cm^2 and mesoporous with an average size of 31 $\pm$ 1.0 nm. Thus, mesoporous nanostructure interface was ideal for realize the electrostatically immobilize between Cyst (negatively charged) and Cyt c (positively charged). The ITO/Fe_3O_4–AuNPs/Cyst bioelectrode showed an improvement in the protein electrochemical behavior, with a 4.0-fold increase in faradaic currents of Cyt c redox processes when compared to standard Cyst-modified polycrystalline gold electrodes [48].

Apart from the aforementioned nanomaterials, nanocomposites are extensively explored for investigations of the direct electrochemistry of Cyt c [49–51]. The researchers' interest is due to the possibility of combining several properties of one or more chemical components to obtain new advanced nanocomposites with improved properties. These advantages emerge from the synergistic effect resulting from supramolecular interactions between (nano)materials (nanoparticles, biomolecules, polymers, organic compounds, inorganic complexes, etc.) [7, 50]. For example, Graphene oxide (GO)-based nanocomposites has been used as electrode nanomaterial due its desirable physicochemical properties that facilitate the efficient Cyt c DET [49]. In this purpose, Cyt c was immobilized on the nanocomposite constituted of graphene oxide-multiwalled carbon nanotube (GO-MWCNT). This nanocomposite was deposited on AuNPs modified glassy carbon electrode and the MWCNT/Au NPs/GCE electrode surface showed an improvement in DET with high surface coverage value (1.82×10^{-9} mol cm^{-2}) and fast electron transfer kinetics (k_{ET} of 3.4 s^{-1}) [49]. In another approach, it was reported the preparation of a multicomponent nanocomposite (MCNC) film constituted of graphene embedded titanium dioxide nanowires (TiO_2(G) NWs) thiol-functionalized polyaniline (PANI(SH)), and gold nanoparticles (Au NPs) [50]. The ultraporous nanostructure obtained was used to immobilize Cyt c on glassy carbon electrode into three-dimensional nanoarchitectured electrode (3DNE). The 3D multicomponent nanoarchitecture demonstrated improvement in the electrochemical properties of Cyt c and in the electrocatalytic activity for detection of nitrite ions (with enhanced sensitivity, low detection limit, high stability, and good reproducibility). Moreover, the Cyt *c*/TiO_2(G) NWs/PANI(SH)–AuNPs/3DNE surface exhibited high k_{ET} value (25.35 s^{-1}), while the surface coverage value was estimated as 1.439×10^{-4} mol cm^{-2}. The synergistic effect of the chemical components improve the physicochemical properties of the 3DNE interface such as enhancing electron transport properties due the connectivity the layer of PANI(SH) on the surface of TiO2(G) NWs [50].

5 Kinetic Modeling

Redox reactions at the protein/electrode interface can be accompanied by several other processes, such as proton transport, conformational transitions, and redox cofactor ligand exchange. Thus, when ET is the limiting step, the oxidation and reduction rates have an apparent ET rate constant (k_{ET}) [15, 52]. In 1979, based on the Butler-Volmer (BV) theory, Laviron published a mathematical treatment to determine the ET rate constants of redox species adsorbed on solid electrodes [53]. This is the most used method to determine the k_{ET} parameter, mainly due to its simplicity, where the only experimental data needed are the overpotentials. However, some authors [54, 55] have demonstrated that BV theory has some restrictions and boundary conditions that may compromise its application in electron transfer kinetics study of immobilized proteins. For instance, in BV formalism, activation energies for the anodic and cathodic processes are assumed as a linear function of overpotential, disregarding the influence of the reorganization energy (λ) in electron transfer rate constants. Furthermore, it disregards contributions to the rate from states in the electrode at any potential other than the applied potential (i.e. Fermi level) [56]. In this sense, Marcus' theory [57–59] is more convenient, once when applied to heterogeneous systems, it overcomes these restrictions. Based on Marcus' equations, Chidsey [54] derived a relation of ET rate constants ($k_{red/oxi}$) with overpotential ($E-E^{0'}$) for the immobilized redox species (Eq. 1). This relation describes the ET rate between a donor and an acceptor over all Fermi levels in the electrode, in which the contribution of each state is weighted according to Fermi–Dirac statistics, once electron transfer can occur to or from any Fermi level (E_i) in the electrode.

$$k_{red/oxi} = \frac{k_{max}}{\sqrt{4\pi\lambda/RT}} \int_{-\infty}^{\infty} \frac{\exp[-[(\lambda \pm F(E - E^{0'}))/RT - x]^2 RT/4\lambda]}{\exp(x) + 1} dx \quad (1)$$

in which E is the applied potential, $E^{0'}$ the formal potential of adsorbed couple and $x = (E - E_i)F/RT$. k_{max} is the limit of rate constant when the overpotential tends to infinity and is given by Eq. 2:

$$k_{max} = \frac{4\pi^2 V_0^2}{N_A h RT} \exp(-\beta r) \quad (2)$$

in which V_0 represents the maximum electronic coupling, β is the decay coefficient and r is the distance between redox centers. $k_{red/oxi}$ can be calculated from Chidsey's equation based on mathematical method described by Armstrong and coworkers [60].

For most cases of redox proteins adsorbed on modified electrodes, the distance between the electrode surface and the protein active site is sufficiently large, and thus ET normally occurs by a non-adiabatic mechanism. Under these conditions, the efficiency of DET is given by the k_{ET}, which exhibits an exponential decay with the ET distance (d), which can be controlled by the number of spacer groups (e.g. the number of CH_2 groups in the alkyl chain length). As an example, Murgida and coworkers, have studied the distance dependence of the ET rates of Cyt c immobilized on Au and Ag electrodes coated with SAMs of pyridine-terminated alkanethiols (PyC_n/C_{n-1}) [61]. In this study, k_{ET} values (at 24.4 °C) of 42 ± 3, 1700 ± 150, 2500 ± 300 and 2400 ± 500 s^{-1} were observed for PyC_{16}/C_{15}, PyC_{12}/C_{11}, PyC_{11}/C_{10} and PyC_6/C_5 SAMs, respectively. These results indicated that effective k_{ET} can be achieved by maintaining the pyridine chains length in the range of 6–12 unites carbon [61]. This dependence may be expressed simply by Eq. 3, derived from Marcus theory [62, 63]:

$$k_{ET} = k^0 \exp[-\beta(d - d_0) \quad (3)$$

In another work, Luz and coworkers [10] investigated the effect of gold nanoparticles (AuNPs) presence at the Cyt c/electrode interface on ET kinetics. Based on cyclic voltammetric experiments and Marcus Theory, the authors demonstrated that incorporation of AuNPs at protein/electrode interface facilitates electron transfer. Applying Chidsey equation, the maximum rate constant for Cyt c on the AuNP modified electrode was approximately 4 times higher as compared to that on the AuNP lacking electrode. Furthermore, exploring the Eq. 3, it was demonstrated a decrease of 20% in the effective ET distance caused by the presence of AuNPs [10].

6 Final Considerations

In this chapter, the various methods applied to immobilize Cyt c, such as physical adsorption, encapsulation or entrapment, covalent binding and layer-by-layer assembly were discussed. Different electrode modifiers have been employed for enzymatic immobilization in order to achieve favorable conditions for Cyt c direct electron transfer. Among them, metallic nanoparticles, carbon-based nanomaterial and nanocomposites have stood out. Several researchers have demonstrated that these nanomaterials create a suitable environment for Cyt c immobilization by increasing the surface coverage and decreasing the distance between iron center and the electrode surface, factors that directly favor the fast electron transfer. In addition, the main strategies to determine the electron transfer kinetic parameters were discussed based on the Marcus theory, since the Butler-Volmer formalism when applied to immobilized proteins systems, presents some restrictions that lead to over or underestimated parameters.

References

1. Gamero-Quijano, A., Herzog, G., Scanlon, M.D.: Bioelectrochemistry of cytochrome c in a closed bipolar electrochemical cell. Electrochem. Commun. **109**, 106600 (2019)
2. Murgida, D.H.: In Situ spectroelectrochemical investigations of electrode-confined electron-transferring proteins and redox enzymes. ACS Omega **6**(5), 3435–3446 (2021)
3. Wang, Y., et al.: Visual distance readout to display the level of energy generation in paper-based biofuel cells: application to enzymatic sensing of glucose. Microchim. Acta **186**(5), 1–9 (2019)
4. Buaki-Sogó, M., et al.: Enzymatic glucose-based bio-batteries: bioenergy to fuel next-generation devices. Top. Curr. Chem. **378**(6), 1–28 (2020)
5. Geiss, A.F., et al.: Engineering the turnover stability of cellobiose dehydrogenase toward long-term bioelectronic applications. ACS Sustain. Chem. Eng. (2021)
6. Page, C.C., et al.: Natural engineering principles of electron tunnelling in biological oxidation–reduction. Nature **402**(6757), 47–52 (1999)
7. Baig, N., Sajid, M., Saleh, T.A.: Recent trends in nanomaterial-modified electrodes for electroanalytical applications. TrAC Trends Anal. Chem. **111**, 47–61 (2019)
8. Rodriguez-Quijada, C., et al.: Physical properties of biomolecules at the nanomaterial interface. J. Phys. Chem. B **122**(11), 2827–2840 (2018)

9. Luz, R.A., Iost, R.M., Crespilho, F.N.: Nanomaterials for biosensors and implantable biodevices. In: Nanobioelectrochemistry, pp. 27–48. Springer (2013)
10. Luz, R.A., Crespilho, F.N.: Gold nanoparticle-mediated electron transfer of cytochrome c on a self-assembled surface. RSC Adv. **6**(67), 62585–62593 (2016)
11. Wu, J.-F., Xu, M.-Q., Zhao, G.-C.: Graphene-based modified electrode for the direct electron transfer of cytochrome c and biosensing. Electrochem. Commun. **12**(1), 175–177 (2010)
12. Holland, J.T., et al.: Engineering of glucose oxidase for direct electron transfer via site-specific gold nanoparticle conjugation. J. Am. Chem. Soc. **133**(48), 19262–19265 (2011)
13. Martins, M.V., et al.: Evidence of short-range electron transfer of a redox enzyme on graphene oxide electrodes. Phys. Chem. Chem. Phys. **16**(33), 17426–17436 (2014)
14. Armstrong, F.A.: Recent developments in dynamic electrochemical studies of adsorbed enzymes and their active sites. Curr. Opin. Chem. Biol. **9**(2), 110–117 (2005)
15. Léger, C., et al.: Enzyme electrokinetics: using protein film voltammetry to investigate redox enzymes and their mechanisms. Biochemistry **42**(29), 8653–8662 (2003)
16. Eddowes, M.J., Hill, H.A.O.: Novel method for the investigation of the electrochemistry of metalloproteins: cytochrome c. J. Chem. Soc. Chem. Commun. **21**, 771b–7772 (1977)
17. Yeh, P., Kuwana, T.: Reversible electrode reaction of cytochrome c. Chem. Lett. **6**(10), 1145–1148 (1977)
18. Armstrong, F.A.: Insights from protein film voltammetry into mechanisms of complex biological electron-transfer reactions. J. Chem. Soc., Dalton Trans. **5**, 661–671 (2002)
19. López-Bernabeu, S., et al.: Direct electron transfer to cytochrome c induced by a conducting polymer. J. Phys. Chem. C **121**(29), 15870–15879 (2017)
20. Manickam, P., et al.: Recent advances in cytochrome c biosensing technologies. Biosens. Bioelectron. **87**, 654–668 (2017)
21. Chen, H., et al.: Fundamentals, applications, and future directions of bioelectrocatalysis. Chem. Rev. **120**(23), 12903–12993 (2020)
22. Melin, F., Hellwig, P.: Redox properties of the membrane proteins from the respiratory chain. Chem. Rev. **120**(18), 10244–10297 (2020)
23. Zuccarello, L., et al.: Electrocatalysis by heme enzymes—applications in biosensing. Catalysts **11**(2), 218 (2021)
24. Gómez-Mingot, M., et al.: Screen printed graphite macroelectrodes for the direct electron transfer of cytochrome c. Analyst **136**(10), 2146–2150 (2011)
25. Wang, G.-X., et al.: Heme plane orientation dependent direct electron transfer of cytochrome c at SAMs/Au electrodes with different wettability. Chem. Commun. **48**(88), 10859–10861 (2012)
26. Aghamiri, Z.S., Mohsennia, M., Rafiee-Pour, H.-A.: Immobilization of cytochrome c and its application as electrochemical biosensors. Talanta **176**, 195–207 (2018)
27. Yagati, A.K., et al.: Electrochemical performance of gold nanoparticle–cytochrome c hybrid interface for H_2O_2 detection. Colloids Surf. B **92**, 161–167 (2012)
28. Sassolas, A., Blum, L.J., Leca-Bouvier, B.D.: Immobilization strategies to develop enzymatic biosensors. Biotechnol. Adv. **30**(3), 489–511 (2012)
29. Lancellotti, L., et al.: Adsorbing surface strongly influences the pseudoperoxidase and nitrite reductase activity of electrode-bound yeast cytochrome c. The effect of hydrophobic immobilization. Bioelectrochemistry **136**, 107628 (2020)
30. Murphy, M., et al.: Electrochemical biosensor for the detection of hydrogen peroxide using cytochrome c covalently immobilized on carboxyl functionalized ionic liquid/multiwalled carbon nanotube hybrid. Appl. Surf. Sci. **492**, 718–725 (2019)
31. Vander Straeten, A., et al.: Protein-based polyelectrolyte multilayers. Adv. Colloid Interface Sci. **280**, 102161 (2020)
32. Zhang, J., et al.: Layer-by-layer assembly for immobilizing enzymes in enzymatic biofuel cells. Sustain. Energy Fuels **4**(1), 68–79 (2020)
33. An, Q., Huang, T., Shi, F.: Covalent layer-by-layer films: chemistry, design, and multidisciplinary applications. Chem. Soc. Rev. **47**(13), 5061–5098 (2018)

34. Lisdat, F.: Trends in the layer-by-layer assembly of redox proteins and enzymes in bioelectrochemistry. Curr. Opin. Electrochem. **5**(1), 165–172 (2017)
35. Petrila, L.-M., et al.: Polyelectrolyte multilayers: an overview on fabrication, properties, and biomedical and environmental applications. Materials **14**(15), 4152 (2021)
36. Temoçin, Z.: Fabrication of a κ-carrageenan-based electroactive cytochrome c multilayer thin film by an electrostatic layer-by-layer assembly. Bioelectrochemistry **129**, 34–41 (2019)
37. Hitaishi, V.P., et al.: Controlling redox enzyme orientation at planar electrodes. Catalysts **8**(5), 192 (2018)
38. Bollella, P., Katz, E.: Enzyme-based biosensors: tackling electron transfer issues. Sensors **20**(12), 3517 (2020)
39. Karimi-Maleh, H., et al.: Metal-based nanoparticles as conductive mediators in electrochemical sensors: a mini review. Curr. Anal. Chem. **15**(2), 136–142 (2019)
40. Koposova, E., et al.: Oleylamine-stabilized gold nanostructures for bioelectronic assembly. Direct electrochemistry of cytochrome c. J. Phys. Chem. C **117**(27), 13944–13951 (2013)
41. Zhao, Y., et al.: Electrochemical biointerface based on electrodeposition AuNPs on 3D graphene aerogel: direct electron transfer of cytochrome c and hydrogen peroxide sensing. J. Electroanal. Chem. **842**, 16–23 (2019)
42. Zhao, D., Li, L., Zhou, J.: Simulation insight into the cytochrome c adsorption on graphene and graphene oxide surfaces. Appl. Surf. Sci. **428**, 825–834 (2018)
43. Shin, J.-H., et al.: Electrochemical H_2O_2 biosensor based on horseradish peroxidase encapsulated protein nanoparticles with reduced graphene oxide-modified gold electrode. Nano Converg. **7**(1), 1–8 (2020)
44. Zhao, H.-Z., et al.: Mechanisms for the direct electron transfer of cytochrome c induced by multi-walled carbon nanotubes. Sensors **12**(8), 10450–10462 (2012)
45. Izzo, M., et al.: Enhancement of direct electron transfer in graphene bioelectrodes containing novel cytochrome c553 variants with optimized heme orientation. Bioelectrochemistry **140**, 107818 (2021)
46. Hou, H., et al.: One-dimensional mesoporous inorganic nanostructures and their applications in energy, sensor, catalysis and adsorption. Progr. Mater. Sci. **113**, 100671 (2020)
47. Huang, S., Lu, M., Wang, L.: Cytochrome c-multiwalled carbon nanotube and cobalt metal organic framework/gold nanoparticle immobilized electrochemical biosensor for nitrite detection. RSC Adv. **11**(1), 501–509 (2021)
48. Melo, A.F., et al.: Electrochemical behavior of cytochrome c immobilized in a magnetically induced mesoporous framework. ChemElectroChem **6**(23), 5802–5809 (2019)
49. Dinesh, B., et al.: Direct electrochemistry of cytochrome c immobilized on a graphene oxide–carbon nanotube composite for picomolar detection of hydrogen peroxide. RSC Adv. **4**(54), 28229–28237 (2014)
50. Shanmugasundaram, K., et al.: Direct electrochemistry of cytochrome c with three-dimensional nanoarchitectured multicomponent composite electrode and nitrite biosensing. Sens. Actuators, B Chem. **228**, 737–747 (2016)
51. Zhang, M., et al.: Direct electrochemistry of cytochrome c immobilized on one dimensional Au nanoparticles functionalized magnetic N-doped carbon nanotubes and its application for the detection of H_2O_2. Sens. Actuators, B Chem. **282**, 85–95 (2019)
52. Murgida, D.H., Hildebrandt, P.: Redox and redox-coupled processes of heme proteins and enzymes at electrochemical interfaces. Phys. Chem. Chem. Phys. **7**(22), 3773–3784 (2005)
53. Laviron, E.: General expression of the linear potential sweep voltammogram in the case of diffusionless electrochemical systems. J. Electroanal. Chem. Interfacial Electrochem. **101**(1), 19–28 (1979)
54. Chidsey, C.E.: Free energy and temperature dependence of electron transfer at the metal-electrolyte interface. Science **251**(4996), 919–922 (1991)
55. Eckermann, A.L., et al.: Electrochemistry of redox-active self-assembled monolayers. Coord. Chem. Rev. **254**(15–16), 1769–1802 (2010)
56. Luz, R.A., et al.: Enzyme biofuel cells: thermodynamics, kinetics and challenges in applicability. ChemElectroChem **1**(11), 1751–1777 (2014)

57. Marcus, R.A.: On the theory of oxidation-reduction reactions involving electron transfer. I. J. Chem. Phys. **24**(5), 966–978 (1956)
58. Marcus, R.A.: Chemical and electrochemical electron-transfer theory. Annu. Rev. Phys. Chem. **15**(1), 155–196 (1964)
59. Silverstein, T.P.: Marcus theory: thermodynamics CAN control the kinetics of electron transfer reactions. J. Chem. Educ. **89**(9), 1159–1167 (2012)
60. Heering, H.A., Hirst, J., Armstrong, F.A.: Interpreting the catalytic voltammetry of electroactive enzymes adsorbed on electrodes. J. Phys. Chem. B **102**(35), 6889–6902 (1998)
61. Yue, H., et al.: On the electron transfer mechanism between cytochrome c and metal electrodes. Evidence for dynamic control at short distances. J. Phys. Chem. B **110**(40), 19906–19913 (2006)
62. Marcus, R.A.: Electron transfer reactions in chemistry. Theory and experiment. Rev. Modern Phys. **65**(3), 599 (1993)
63. Park, W.-C., Hong, H.-G.: Determination of reorganization energy from the temperature dependence of electron transfer rate constant for hydroquinone-tethered self-assembled monolayers (SAMs). Bull. Korean Chem. Soc. **27**(3), 381–385 (2006)

Surface-Confined Biomolecules for Application in Bioelectronics

Rodrigo M. Iost

1 Immobilization of Biomolecules

The spatial confinement of biomolecules is usually prepared intentionally aiming to obtain practical advantages over enzymes in aqueous solution, such as easy of separation of the biocatalyst from the products of reaction. The possibility to recover the biocatalyst in the final of its use is also experimentally desirable [1]. In particular, it is necessary to ensure that the biomolecules remains at least partially active after the immobilization [2]. To maintain, at least partially, the stability and biocatalytic activity of enzymes is also a prerequisite for practical application of biosensors [3–6], bioenergy conversion [7, 8] or bioelectro production of chemicals [9–13] for industry. In the course of the last decades, numerous applications regards to the use confinement of biomolecules has increased considerably, either to a surface or restricting them to a volume to which they are confined [14, 15]. However, to bring biomolecules to non-natural physical or chemical environments is still a difficult task, especially when the maintainance of biocatalytic properties are desirable. We can briefly emphasize the basic protocols of immobilization of biomolecules [8, 14, 16]: (i) the physical adsorption; (ii) entrapment; (iii) covalent binding and; (iv) encapsulation. Figure 1 shows a scheme of some basic methods of immobilization of biological molecules, such as proteins and enzymes.

The immersion of a solid substrate in to a solution that contains the biological material can be easily used to attach them on surface. The physical adsorption of the biomolecules (Fig. 1a) to a surface is feasible basically due to surface interactions of van der Waals forces, electrostatic forces, hydrogen bonds and/or hydrophobic interactions [17]. From the experimental point of view, the physical adsorption of biological molecules is relatively simple and can be carry out by the immersion of a solid substrate in to a solution that contains the biological material [18] or by dropping

R. M. Iost (✉)
Instituto de Química de São Carlos, Universidade de São Paulo, São Carlos, SP 13560-970, Brazil
e-mail: rodrigoiost@usp.br

F. N. Crespilho (ed.), *Advances in Bioelectrochemistry Volume 1*,
https://doi.org/10.1007/978-3-030-94988-4_4

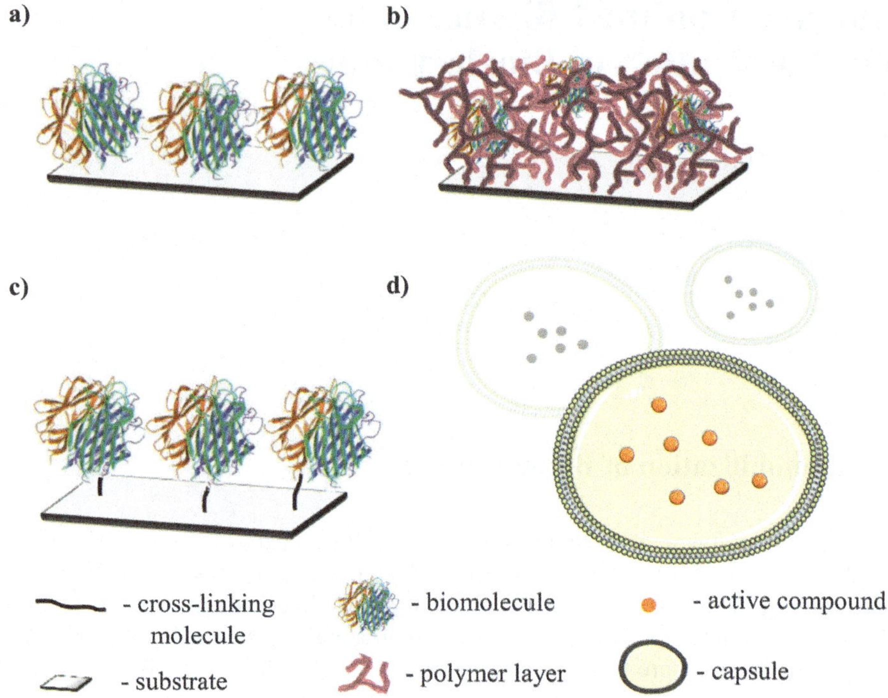

Fig. 1 Immobilization of Biomolecules. **a** physical adsorption, **b** entrapment in polymer matrix, **c** covalent attachment and **d** encapsulation

a solution of proteins on to the solid surface [19–26], previously chemically modified or not. Soft or flat surfaces of metals, semiconductors, and some carbon nanomaterials like graphene monolayer are relatively difficult to maintain the biomolecules physically adsorbed by itself and they can be easily leached to the bulk solution. Concerning the application in biodevices, it is desirable that the redox centers of biomolecules are efficiently connected to the surface at the same time favoring the efficiency of biodevice's characteristics. For example, this is directly related to the concentration of the substrate used to obtain an efficient response in a biosensor [27, 28] and/or also to achieve the best performance regards to the kinetic and thermodynamic parameters of (bio)energy conversion devices [29, 30]. The synthesis of highly porous materials has been extensively used to circumvent this experimental hurdle and generally to improve the characteristics of the device. One of the most extensively used porous materials for biodevices application is carbon [29, 31–33]. Highly porous carbon can be easily obtained experimentally through oxidative or non-oxidative protocols and are reported to obtain an efficient catalytic performance of the biodevice [34, 35]. Also, porous materials are defined in relation to its size and eventually can increase the feasibility to which biomolecules are connected with the surface [36]. Besides the simple adsorption of biomolecules on flat surfaces or highly porous materials, it is also crucial to understand the role played by the presence of chemical functionalities on solid surfaces.

The presence of chemical functionalities plays a pivotal role for the fabrication of efficient biodevices [37]. One of the most efficient methodologies to modify and modulate the chemistry characteristics of surfaces is by the use of self-assembly of organic/inorganic materials by using layer-by-layer technique [38] or even the use of Langmuir–Blodgett [39, 40] films. These protocols are extensively reported to modify the surface of electrodes previously to the adsorption of biological molecules, and enables the maintenance of the structural integrity of the biomolecules at the same time can control the thickness of the coating layer [41]. Another efficient protocol to confine biomolecules on surfaces is by the use of entrapment (Fig. 1b). Entrapment, or occlusion of biomolecules, involves the use of a matrix (e.g. polymer solution) for dispersion biological molecules [42, 43]. The solution containing the biological material is then used to coat the surface by drop-casting, or immersion of the solid substrate in to the solution. For this purpose, organic polymers and/or polyelectrolytes are dispersed with the biomolecules in solution. The possibility to bound biomolecules on surface have also been subject of major interest for bioelectronics, such as by using covalent [44–46] or non-covalent [47] protocols of functionalisation of surfaces. Covalent or even non-covalent modifications of surfaces are one of the most extensively explored protocols for tunning the chemistry of surfaces, also due to its experimental simplicity and easy of control of the thickness of the coating layer. Another important advantage of using functionalised surfaces is the orientation of the biomolecule relative to the solid surface [48], an important characteristic when efficient biorecognition platforms are desirable [48, 49].

Covalent modification of surfaces (Fig. 1c) can be advantageous over non-covalent functionalisation because they can drive the way how biomolecules are oriented in relation to the surface [50]. One of the most exployed protocols for covalent functionalisation of surfaces is by the grafting of aryldiazonium salts on to solid supports [51]. The use of diazonium salt reaction is interesting due to some main characteristics. For example, many different diazonium salts are commercially available or can be easily synthesize, the functionalisation can be carry out on conducting or non-conducting surfaces and the modification is very stable due to the formation of a covalent bound between the surface and the aryl chemical groups [52]. The two common methods for the attachment of aryldiazonium salts on surfaces are by the use of thermal [53] or the electrochemical [54] modification, and can be then easily probed by Raman [53–56] or infrared [57, 58] spectroscopy. The use of cross-linking molecules has also been reported to anchor a wide variety of biomolecules on surfaces for biosensing and bioenergy conversion applications. Among the many available cross-linking molecules, glutaraldehyde is undoubtedly one of the most reported in the last decades for the immobilization of biomolecules on surface [59]. Glutaraldehyde is a linear dialdehyde and high soluble molecule in water/alcohol and in some organic solvents [59]. It is also reported that the advantage of using such type of molecule is the relative low cost and its commercial availability. The reactivity of the molecule with amino chemical functionalities of the side chains of enzymes generates highly stable chemical compounds with amino-functionalized surfaces, favoring the attachment of proteins [60, 61]. Finally, the encapsulation of biological molecules are also used to maintain their integrity and structural stability.

Capsules has also been usually used for creation of micro- or nanoscale reservoir-like structures with a relatively defined coating material. Basically, the microencapsulation of molecules, or active chemical compounds (core), is defined by the creation of a physical barrier (shell) to restrict the way how the core are in contact with the external media aiming to control the reactivity of the inner environment, to control the release of the core material, etc. [62, 63]. The core or internal phase of the material can be solid, liquid or gas, and the shell is usually prepared from porous polymeric materials aiming practical applications [64]. The morphology of a capsule can also be defined by its simple or multiple walls of the shell, or also according to the matrix used to obtain the capsules. The simplest form of a capsule is a relative small spheres with a uniform wall around it. Polymers are usually used for creation of these capsules due to the possibility to control the characteristics of porosity for applications in food and target drug-release [65, 66]. On the other hand, a range of lipid types has been used as model membranes of cell-size to form "giant" unilamellar vesicles (Fig. 1d) for practical applications, e.g., as drug delivery synthetic systems.

Among these experimental protocols for immobilization of biomolecules, it is interesting to understand and be able to probe the interface between the biomolecule and the solid surface. Regards to the chemical functionalization protocols for the fabrication of biodevices, it is experimentally interesting to ensure the feasibility of the protocol for the functionalization of surfaces, especially when multiple steps of chemical modification are realized previously to the immobilization of the biomolecules. However, to probe chemical functionalities can be experimentally difficult. Strictly speaking, appropriate and feasible protocols of immobilization needs to guarantee their efficiency by probing the presence of target groups on surface or by its presence in the molecular structure of the biomolecule. In the next topics, we will focus on the description of the protocols and strategies used for the immobilization of biomolecules regards to the probing of chemical functionalities on to solid surfaces for application in bioelectronics. We will also focus on the electrochemistry and spectroscopic techniques used for the characterization of chemical functionalities for immobilization of biomolecules on surfaces. The site-selective chemical modification aminoacid side chains of proteins [67, 68], cascade reactions [69, 70], the effect of molecular crowding [71] phenomena of biological molecules are equally important and can be found in specific reviews.

2 Surface-Confined Biomolecules—Chemical Functionalisation

The covalent or non-covalent chemical functionalisation of micro- and nanomaterials can play a pivotal role for practical application in many fields of research, especially when few quantities of biomolecules are studied. For example, the chemical functionalisation of solid surfaces are extensively reported for the immobilization of many types of biomolecules such as DNA [72], redox proteins and enzymes [73], peptides [74] and virus [75, 76]. The advantage of such protocols is the relative stability of the bounds involving the biomolecules and the substrate, favoring that the biomolecules do not release completely in to the bulk solution or lose the biological activity during the practical application. For this purpose, many types of precursors are used aiming to achieve the attachment of chemical groups on to carbon, metallic and semiconducting substrates [77], and can be systematically probed by the vibrational fingerprint of the chemical functionalities by using vibrational spectroscopy techniques.

Carbon nanomaterials (e.g. nanotubes [37, 78], functionalized graphene flakes [79–81] or graphene monolayer [82]) and metal surfaces (e.g. gold electrodes [83–86]) are extensively used as platforms for biosensing. It is reported that both covalent and non-covalent functionalisation of carbon surfaces at nanoscale or even at atomic scale provides some advantages or disadvantages when they are used as platforms for subsequent immobilization of biomolecules [47]. As an example, the covalent modification of surfaces can strongly attach the biomolecules on surfaces previously modified with organic moities, controlling the charge characteristics on surface, improving the efficiency of immobilization and increasing the stability of the biodevice [47]. On the other hand, the attachment of covalent moities can alter the electronic structure of nanomaterials and hampers the fabrication of the biodevice [87, 88]. In the next topics we will emphasize the capability to probe chemical functionalities at nano and atomic scale level aiming the application in the attachment of biomolecules for bioelectronics. Moreover, it will be emphasize some main protocols for modification of surfaces using organic molecules to control the chemistry characteristic at the interface by the use of galvanostatic or potentiostatic deposition [89], and using the "click-chemistry" of molecules on surfaces [90].

3 Non-covalent Chemical Functionalisation

The non-covalent binding energy is fundamentally conceived as lower in comparison to covalent bonds, and can contribute for the attraction or cohesion in physical and chemical systems [91]. In regards to the non-covalent functionalisation of surfaces, it is desirable to identify the fingerprint of chemical functionalities on surface for subsequent immobilization of biomolecules, especially when lower concentrations of the precursor are used. When micro- and nanoscale surfaces are designed, it is desirable to develop simple methods for probing chemical functionalities on surfaces. Such protocols involves basically the modification of surfaces with the chemical functionalities and the vibrational fingerprint of chemical functionalities can be then probed spectroscopically by Raman [55, 92–94] or infrared spectroscopy [95–98].

The probing of chemical functionalities on carbon nanomaterials has received considerable attention in the past decades. For example, the chemical functionalization plays a pivotal role for modulation of the chemistry characteristics of solid conducting surfaces, e.g., gold [99], carbon nanotubes [44, 100] or graphene monolayer [32, 56, 73, 101] electrodes. For this purpose, some studies has focused on the role of modification at the basal plane and at the edge of micro- and nanoscale electrodes and their local electrochemistry influence in the final properties of the device. Also, the probing of functional chemical groups in isolated carbon nanotubes and graphene is currently investigated with focus on the development of more sensitive and selective sensing and biosensing devices. As an example of non-covalent chemical functionalisation of isolated carbon nanomaterials, poly(pyrrole) (PPy) was used to tune the surface properties of single carbon nanotubes and graphene monolayer electrodes using electrochemistry [55]. Atomic Force Microscopy (AFM) was used to verify the relative height of the thickness of the polymer layer deposited on to the nanoscale electrodes, as shown by the relative height of around 2 nm for graphene and 3 nm for single carbon nanotubes. The relative height for AuNPs was in the range from 20 to 50 nm. The Raman spectrum for both single carbon nanotubes and bare graphene electrodes showed the characteristic D-, G- and 2D peaks that in general does not effect the Raman signature of the underlying single nanomaterial, before and after the local probe of the chemical funcionalisation [55]. In this case, the strenght of near-field enhancement effect caused by the presence of AuNPs strongly influences the observation of the chemical functionalisation, and basically was used to assign the various peaks related to the C–H bending, C–N stretching and C=C stretching and pyrrole ring stretching modes of polymer PPy. The spatial distribution of these chemical fingerprints on graphene monolayer were also monitored by Raman [55]. In another approach, the influence of chemical functionalities electrochemically deposited on to the surface of graphene monolayer on surface charge was experimentally and theoretically investigated for future applications in sensing and biosensing devices [101]. For this purpose, the strategy of modifying the chemical functionalisation of graphene surface was used to modulate the isoelectric point (pI) that is related to the acid–base properties of the ionizable chemical groups on surface. The possibility to modulate the chemistry characteristics of the isolated edge of carbon nanomaterials can also be experimentally advantageous over the basal plane of electrodes.

The possibility to control the chemistry at the periphery or at an isolated edge of graphene monolayer electrodes has received considerable attention aiming to control its interfacial physical and chemistry characteristics [102–106]. Some fundamental studies have reported that the reactivity of electrodes at the periphery or even at isolated edge of graphene monolayer can change considerably when compared to bulk carbon electrodes [106–108]. The funcionalisation of graphene monolayer occurs at both basal plane or at the periphery of the graphene edge, especially when graphene flakes are dispersed in solution [109, 110]. Then, it is crucial to use experimental protocols for efficient probing of such chemical functionalities on surfaces, especially at the atomic scale edge of carbon materials. This is because atomic scale surfaces are envisioned for application in the monitoring of the activity of molecules in biodevices.

On the other hand, a simple and selective method for non-covalent functionalisation of na isolated graphene edge was reported in a recent study by using electrochemistry [111], as shown in Fig. 2.

The fabrication procedure for the functionalisation of the isolated graphene edge is described as follows. Graphene monolayer samples obtained by chemical vapor deposition (CVD-grown graphene) method and transferred to Si/SiO_2 substrates with pre-patterned metallic Ti/Pt contacts by using wet-transfer protocol [112]. The wet-transfer protocol envolves the drop of poly(styrene) solution in toluene over graphene as a support polymer [113]. The transfer of CVD-grown graphene was followed by the removal of trace metal/metal oxide copper particles and remaining organic polymer from the wet-transfer protocol from the surface of graphene monolayer by electrochemical etching (*e-etching*) [112]. A positive photoresist was then deposited over the surface of graphene samples to define the location of graphene edge (GrEdge) by exposure of the samples to oxygen plasma. GrEdge was subsequently modified with poly(aminobenzylamine) (pABA) non-covalently attached

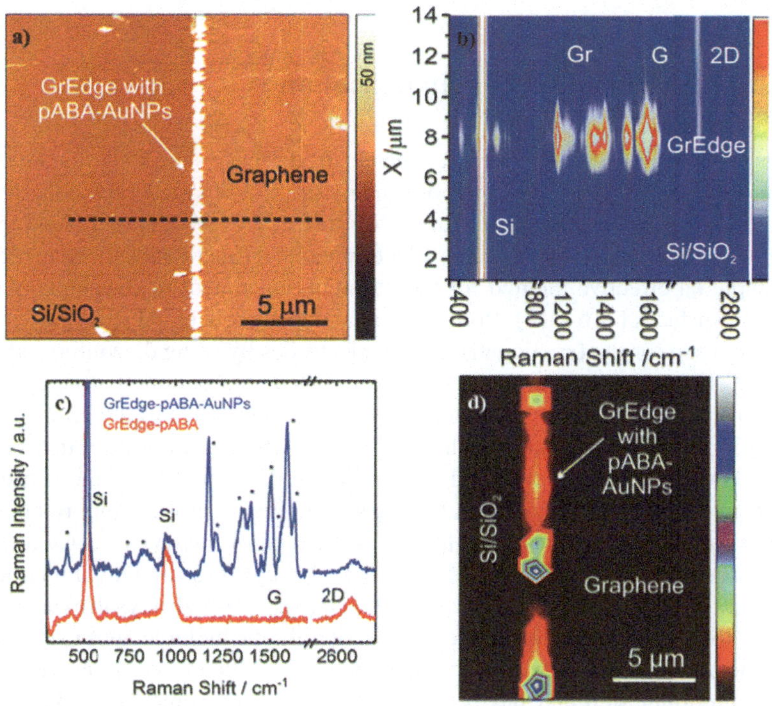

Fig. 2 Probing an isolated graphene edge. **a** AFM image of a typical GrEdge functionalized with poly(aminobenzylamine) (pABA) and AuNPs, referred to as GrEdge–pABA–AuNPs. **b** Comparison of the Raman spectra at GrEdge–pABA (red curve) and GrEdge–pABA–AuNPs (blue curve). **c** Raman map along the line shown in (**a**). **d** Map of the C–N stretching mode (1358 cm^{-1}) of the same region as in (**a**) (λ_{ex}: 633 nm, 3.4 mW, 2 × 1 s). Reproduced with permission of Ref. [111]. Copyright 2019, Royal Society of Chemistry

exclusively at isolated edge of graphene, with and without the electrodeposition of AuNPs [111]. In this case, the covalent attachment through diazonium coupling reaction was also carry out, with no possibility of subsequent deposition of AuNPs at GrEdge. The deposition of the AuNPs was then carried out in order to characterize spectroscopically the non-covalent attachment of chemical functionalities especifically attached to the edge of graphene monolayer by SERS effect. An AFM image of a typical GrEdge-pABA-AuNPs is shows in Fig. 2a. Figure 2b shows the Raman spectra at the functionalised GrEdge before (red) and after (blue) attached to AuNPs, with new Raman peaks in addition to the D, G and 2D peaks of graphene and related to the pABA. The Raman modes are assigned to the specific modes of the polymer pABA (Fig. 2c). The Raman map along the cross-section of GrEdge is also shown in Fig. 1d with the corresponding spots related to the vibrational modes of pABA layer electrochemically deposited at GrEdge-AuNPs [111]. The possibility to probe such chemical functionalities at isolated graphene edge also opens the pathway for obtention of more sensitive sensing devices [114] and future immobilization of biomolecules exclusively at the edge.

4 Covalent Chemical Functionalisation

The preparation of covalent functionalized electrodes have shown significant advantage in bioelectronics. For this purpose, many are the reported protocols that can generate covalent chemical functionalities on surfaces, especially on gold [115] and carbon [116] materials. The thermal or electrochemical functionalisation are reported as a simple an effective method for the covalent attachment of oxygen containing groups on surfaces [116–118]. Both thermal and electrochemical strategies are also frequently used for the functionalization of electrodes by using diazonium salt reaction [119, 120]. Diazonium reaction has been used as a common strategy for the control of the attachment of covalent moities on surfaces due to the efficiency and simplicity of the experimental methodology [51]. The main question that remains is the side chemical group exposed to the surface that could favor the subsequent attachment [72, 121]. The nature of the organic moities is also an important experimental parameter to be considered for the fabrication of biodevices, such as by the use of thionyl chloride ($SOCl_2$) [122, 123] or 1-ethyl-3-(3-dimethylaminopropyl) carbodiimide (EDC) N-hydroxysuccinimide (NHS), or N-hydroxysulfoxuccinimide (sulfo-NHS) for the formation amide bonds that links the biomolecules to the surface of the electrode [124]. The characterization of the efficient coupling reaction is also an important step and has been reported in some promissor studies aiming the fabrication of biodevices, as shown for the functionalization of gold electrodes for immobilization of redox proteins (Fig. 3).

It is reported that the coupling strenght of biomolecules on surface as well as their relative orientation plays a pivotal role for mechanistic investigation of charge transport in protein electronics. One of the most relevant post-functionalization method for chemical modification and coupling proteins, DNA and virus on surfaces is the use

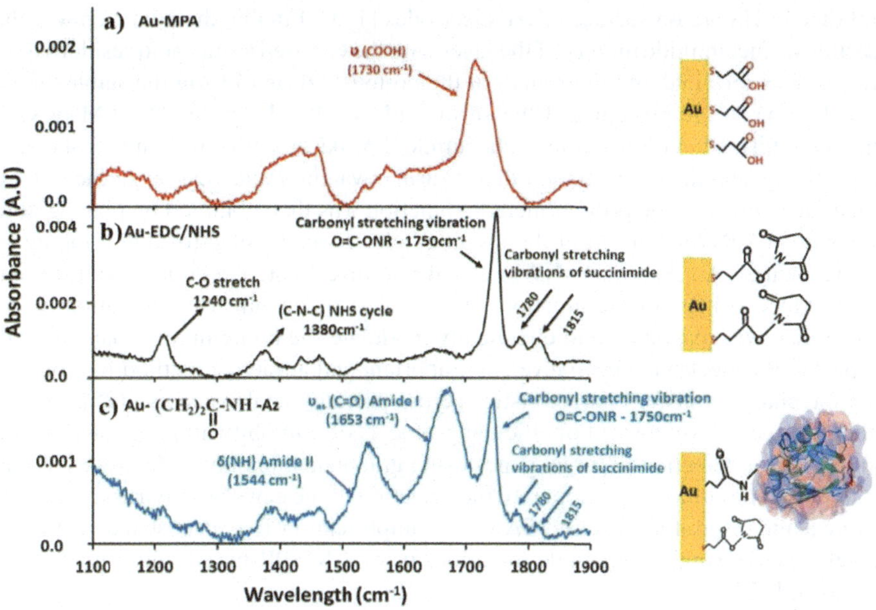

Fig. 3 Protein electronics. PM-IRRAS spectra in the 1100 cm^{-1} and 1900 cm^{-1} range measured on **a** monolayer of MPA bound to the Au surface, main peak at 1730 cm^{-1} of the (C=O) stretching vibration of free carboxylic (–COOH) group. **b** MPA monolayer on Au surface following reaction with EDC/NHS. **c** EDC/NHS activated MPA monolayer on Au surface following the reaction with the protein Azurin forming the amide I and amide II peaks at 1653 cm^{-1} and 1544 cm^{-1}, respectively. Reproduced with permission of Ref. [125]. Copyright 2018, American Chemical Society

of EDC as coupling agents to carboxyl group functionality [124]. The amine-reactive ester is then stabilised by the use of NHS or sulfo-NHS before the coupling reaction with the biomolecules, thus increasing the efficiency of EDC/NHS coupling reaction [124]. The EDC/NHS coupling reaction has been recently shown to covalently attach the protein azurin to gold contacts as well as to modulate the energy level alignment of protein/gold junctions [125]. Figure 3a shows the polarization modulation-infrared reflection-adsorption spectroscopy (PM-IRRAS) spectra of the layer of mercapto-propionic acid (MPA) deposited on surface of Au electrodes. The spectral range was monitored from 1100 to 1900 cm^{-1}. The MPA bound to the surface of the electrode shows one main vibrational stretching peak at 1730 cm^{-1} related to (C=O) bounds of carboxylic acid (COOH). Then, the MPA layer was reacted with EDC, followed by the reaction with sulfo-NHS (Fig. 3b). The second step of functionalisation revealed the replacement (shift) of 1730 cm^{-1} peak by the peak at 1750 cm^{-1}, along with two other peaks centered at 1780 cm^{-1} and 1815 cm^{-1} that corresponds to carbonyl and succinimide stretching vibrations, respectively [125]. The realization of the next step of chemical modification was carry out using lower concentration of EDC/NHS in order to chemically activated the free –COOH functionalities of the MPA layer. This step of modification was also supported by PM-IRRAS spectra where both –COOH and EDC-NHS are on surface of Au electrodes [125]. Finally, the next step was the reaction of succinimide moiety of the layer with the exposed amino acid residues (e.g. lysines or N-terminal amino groups) of the protein Azurin to form the stable amide bound between the protein and the surface of the Au electrode. The PM-IRRAS shows the formation of the amide I and amide II peaks at 1653 cm^{-1} and 1544 cm^{-1}, respectively, and the confirmation that Azurin covalently attached on surface of Au electrode (Fig. 3c). The protein/metallic junction was then obtained by placing gold nanowires (AuNWs) on top of the modified electrodes by electrostatic trapping of the nanowires (Au-Azurin-AuNWs) in order to investigate the current–voltage and conductance-voltage of the junction. In this case, the chemical modification of the Au surface was responsible to chemically modulate the tunneling mechanism with respect to the electrode Fermi level, an important and attractive method for controlling the charge-transfer characteristics across protein/metal contacts [125]. On the other hand, the development of efficient protocols of immobilization has also shown promissor for detection of target genes using graphene monolayer electrodes. As an example, some authors have reported the detection of the new coronavirus that causes severe acute respiratory disease (SARS) by amplification-free electrical detection of target genes or single-nucleotide mutations using CRISPR-based liquid-field effect transistor [126].

The capability to monitor the biomolecular interactions with single-molecule sensitivity is still a experimental challenge due to the difficulty to monitor the dynamics of biomolecular interactions at single entity level. A relatively recent protocol related to a cathegory of synthesis was introduce by Sharpless and co-workers to describe modular and stereospecific reactions. The advantage of such protocols is to obtain site-specific reactions with possible control of molecular orientations towards single biomolecule detection approach. This is a key aspect for future nanoelectronics and the process of fabrication is also dependent on how single nanomaterials are manipulated. In a recent study, Freeley and co-workers reported for the first time a site-specific "one-click" covalent coupling of single proteins on tip-end of individual carbon nanotubes with envisioned application in bioelectronics [127] (Fig. 4).

Figure 4a shows the attachment process scheme of a single protein at the tip-end of individual carbon nanotubes. For this purpose, a single-stranded DNA (ssDNA) allowed the dispersion of the carbon nanotubes in aqueous solution. The strategy was the wrapping of the sidewalls of the nanotubes by ssDNA through π–π stacking, leaving the terminal end of the nanotubes available for the amidation reaction on the exposed –COOH functionality. The reaction at the tip end of the individual carbon nanotubes was carry out with dibenzocyclooctyne (DBCO-functionalized ssDNA/SWCNTs) and subsequent reacted with azide groups promoted 1,3-dipolar cycloaddition [127]. Two engineered green fluorescente proteins (sfGFP) variants was then attached to the end of the nanotubes and shows the importance of linkage site for the protein-SWCNTs communication [127]. Basically, the generation of short axis GFP (GFP^{SA}) mutants to be in proximity with the end of the tubes was realized by substitution of Gln204 by Glu132, with minimal impact in the biological

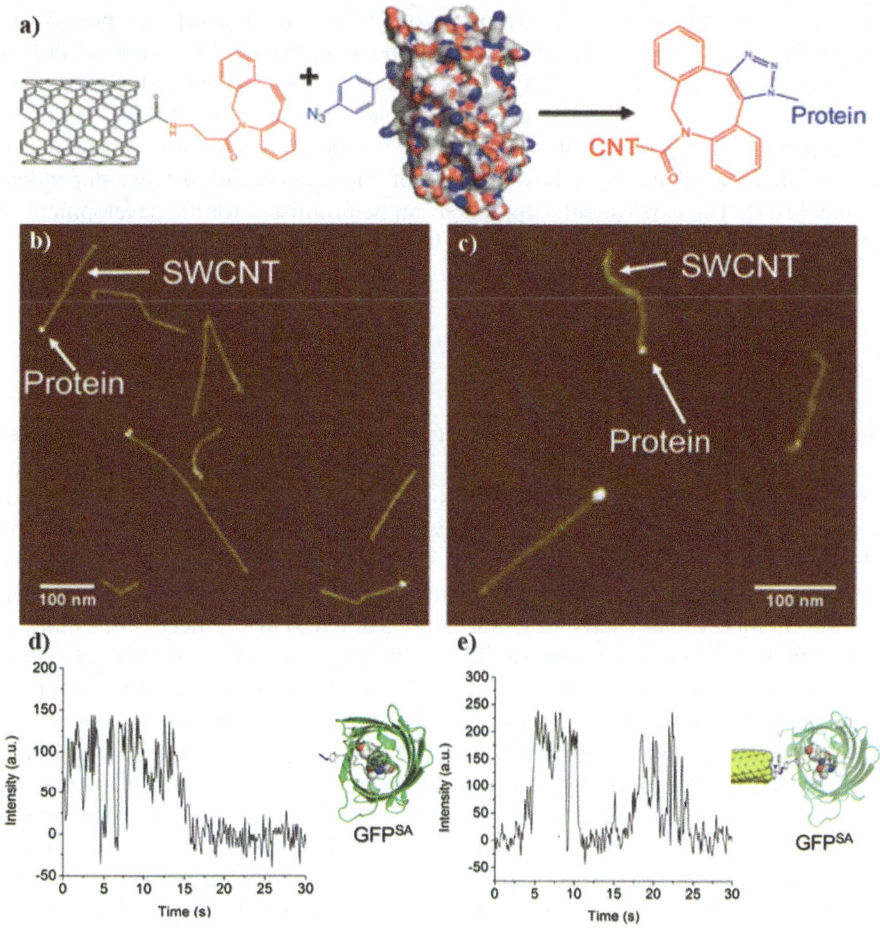

Fig. 4 SWCNT Biodevice. **a** Schematics of the single protein attachment to SWCNTs. **b** and **c** shows AFM image of ssDNA/SWCNT-GFPLA hybrids. Z-scales = 6 nm. Representative single-molecule fluorescence traces for **d** GFPSA and **e** SWCNT-GFPSA hybrid. Reproduced with permission of Ref. [127]. Copyright 2017, American Chemical Society

functions of GFP. The GFP mutants with modified amino acid residues was viabilized to be attached to the end of individual carbon nanotubes in aqueous solution ssDNA/SWCNT-GFP hybrids [127]. Figure 4b and c shows the AFM imaging of the substrate and the prove of attachment of the proteins to the tip end of the nanotubes. The AFM images corresponding to the relative height profiles of the GFP at the tip end of the nanotubes of around 2.5 nm. Figure 4d and e shows the "on" and "off" states by monitoring of GFP-nanotube tip. The nanohybrids were monitored with single-molecule resolution by dispersion of ssDNA/SWCNT-GFP and spaced at least 1 μm apart on glass substrates. The measurements were carry out by comparison of total internal reflectance fluorescence microscopy of the GFP and the nanohybrids.

The single molecule fluorescence dynamics of GFP showed that off-time periods are independent of excitation intensity [127]. On the other hand, GFP^{SA} showed shorter off-times (between "on" and "off" states), while the nanohybrids showed longer off-time periods. Also, the relatively short distance between the GFP and the tip of the carbon nanotubes reveals that the electronic coupling occurs via photo-induced charge transfer mechanism, or the quenching of fluorescence in a distance-dependent manner [127]. The experimental approach can be promissor for the development of solution processable single-molecule bioelectronics.

5 Final Considerations

The confinement of biomolecules on surfaces has developed as a consequence of its scientific and technological application in bioelectronics. An inquirity about the pivotal role of immobilized biomolecules may provide some technological advantages regards to, e.g., their biological activity and stability. This is the main reason why the control of chemistry characteristics of surfaces are experimentally attractive for application in bioelectronics, and the use of experimental protocols for controlling the chemistry characteristics of surfaces have led to many successful examples to the study of biomolecules on surfaces. The accurate control of chemical functionalisation can be used to tune the chemistry characteristics at the interface, either by non-covalent or covalent, and can represent a strategic pathway to improve the fabrication of biosensors and bioelectronics. From this point of view, it is useful to develop simple and promissor strategies to probe molecular events at micro- and nanoscale level, e.g., using electrochemistry or electronic/vibrational spectroscopy. Although much progress has been made in order to identify the chemistry characteristics on surfaces or biological events using few quantities of biological molecules is still a challenge to be overcomed. Finally, which are the most efficient method of immobilization could be decided aiming to a specific technological application and will also depend on the basic research and the capability of investigation of interfacial characteristics of micro- and nanoscale systems.

References

1. Tischer, W., Wedekind, F.: Immobilized Enzymes: Methods and Applications (1999)
2. Crespilho, F.N., Iost, R.M., Travain, S.A., Oliveira, O.N., Zucolotto, V.: Enzyme immobilization on Ag nanoparticles/polyaniline nanocomposites. Biosens. Bioelectron. (2009)
3. Iost, R.M., Silva, W.C. Da, Madurro, J.M., Madurro, A.G.B., Ferreira, L.F., Crespilho, F.N.: Recent advances in nano-based electrochemical biosensors: application in diagnosis and monitoring of diseases. Front. Biosci. Elit. (2011)
4. Iost, R.M., Madurro, J.M., Brito-Madurro, A.G., Nantes, I.L., Caseli, L., Crespilho, F.N.: Strategies of nano-manipulation for application in electrochemical biosensors. Int. J. Electrochem. Sci. (2011)

5. Iost, R.M., Crespilho, F.N.: Layer-by-layer self-assembly and electrochemistry: applications in biosensing and bioelectronics. Biosens. Bioelectron. (2012)
6. Siqueira, J.R., Caseli, L., Crespilho, F.N., Zucolotto, V., Oliveira, O.N.: Immobilization of biomolecules on nanostructured films for biosensing. Biosens. Bioelectron. (2010)
7. Luz, R.A.S., Pereira, A.R., Iost, R.M., Crespilho, F.N.: Biofuel cells. In: Nanoenergy: Nanotechnology Applied for Energy Production, 2nd edn. (2017)
8. Luz, R.A.S., Pereira, A.R., de Souza, J.C.P., Sales, F.C.P.F., Crespilho, F.N.: Enzyme biofuel cells: thermodynamics, kinetics and challenges in applicability. ChemElectroChem (2014)
9. Hartman, T., Wondergem, C.S., Kumar, N., Van Den Berg, A., Weckhuysen, B.M.: Surface- and tip-enhanced Raman spectroscopy in catalysis. J. Phys. Chem. Lett. (2016)
10. Milton, R.D., Cai, R., Abdellaoui, S., Leech, D., De Lacey, A.L., Pita, M., Minteer, S.D.: Bioelectrochemical Haber–Bosch process: an ammonia-producing H_2/N_2 fuel cell. Angew. Chem. Int. Ed. (2017)
11. Abdellaoui, S., Macazo, F.C., Cai, R., De Lacey, A.L., Pita, M., Minteer, S.D.: Enzymatic electrosynthesis of alkanes by bioelectrocatalytic decarbonylation of fatty aldehydes. Angew. Chem. Int. Ed. (2018)
12. Yuan, M., Sahin, S., Cai, R., Abdellaoui, S., Hickey, D.P., Minteer, S.D., Milton, R.D.: Creating a low-potential redox polymer for efficient electroenzymatic CO_2 reduction. Angew. Chem. (2018)
13. Varfolomeyev, S.D.: Bioelectrosynthesis as an alternative to photosynthesis. Appl. Biochem. Biotechnol. Part A Enzym. Eng. Biotechnol. (1992)
14. Küchler, A., Yoshimoto, M., Luginbühl, S., Mavelli, F., Walde, P.: Enzymatic reactions in confined environments. Nat. Nanotechnol. (2016)
15. Liebherr, R.B., Gorris, H.H.: Enzyme molecules in solitary confinement. Molecules (2014)
16. Crespilho, F.N.: Nanobioelectrochemistry: from implantable biosensors to green power generation (2013)
17. Fritz, P.A., Bera, B., Van Den Berg, J., Visser, I., Kleijn, J.M., Boom, R.M., Schroën, C.G.P.H.: Electrode surface potential-driven protein adsorption and desorption through modulation of electrostatic, van Der Waals, and hydration interactions. Langmuir (2021)
18. Schmidt, T.F., Caseli, L., dos Santos, D.S., Oliveira, O.N.: Enzyme activity of horseradish peroxidase immobilized in Chitosan matrices in alternated layers. Mater. Sci. Eng. C (2009)
19. Caseli, L., Siqueira, J.R.: High enzymatic activity preservation with carbon nanotubes incorporated in urease-lipid hybrid Langmuir-Blodgett films. Langmuir (2012)
20. Schmidt, T.F., Caseli, L., Viitala, T., Oliveira, O.N.: Enhanced activity of horseradish peroxidase in Langmuir-Blodgett films of phospholipids. Biochim. Biophys. Acta Biomembr. (2008)
21. Caseli, L., dos Santos, D.S., Aroca, R.F., Oliveira, O.N.: Controlled fabrication of gold nanoparticles biomediated by glucose oxidase immobilized on Chitosan layer-by-layer films. Mater. Sci. Eng. C (2009)
22. Schmidt, T.F., Pavinatto, F.J., Caseli, L., Gonzaga, M.L.C., Soares, S.A., Ricardo, N.M.P.S., Oliveira, O.N.: Interaction of polysaccharide-protein complex from agaricus Blazei with Langmuir and Langmuir-Blodgett films of phospholipids. J. Colloid Interface Sci. (2009)
23. Goto, T.E., Lopez, R.F., Oliveira, O.N., Caseli, L.: Enzyme activity of catalase immobilized in Langmuir-Blodgett films of phospholipids. Langmuir (2010)
24. Caseli, L., Crespilho, F.N., Nobre, T.M., Zaniquelli, M.E.D., Zucolotto, V., Oliveira, O.N.: Using phospholipid Langmuir and Langmuir-Blodgett films as matrix for urease immobilization. J. Colloid Interface Sci. (2008)
25. Crespilho, F.N., Lanfredi, A.J.C., Leite, E.R., Chiquito, A.J.: Development of individual semiconductor nanowire for bioelectrochemical device at low overpotential conditions. Electrochem. Commun. (2009)
26. Pereira, A.R., de Souza, J.C.P., Iost, R.M., Sales, F.C.P.F., Crespilho, F.N.: Application of carbon fibers to flexible enzyme electrodes. J. Electroanal. Chem. (2016)
27. Hoeben, F.J.M., Meijer, F.S., Dekker, C., Albracht, S.P.J., Heering, H.A., Lemay, S.G.: Toward single-enzyme molecule electrochemistry: [NiFe]—hydrogenase protein film voltammetry at nanoelectrodes. ACS Nano (2008)

28. Willner, I., Katz, E.: Integration of layered redox proteins and conductive supports for bioelectronic applications. Angew. Chem. Int. Ed. (2000)
29. Gao, F., Viry, L., Maugey, M., Poulin, P., Mano, N.: Engineering hybrid nanotube wires for high-power biofuel cells. Nat. Commun. (2010)
30. Zebda, A., Gondran, C., Le Goff, A., Holzinger, M., Cinquin, P., Cosnier, S.: Mediatorless high-power glucose biofuel cells based on compressed carbon nanotube-enzyme electrodes. Nat. Commun. (2011)
31. Deng, X., Li, J., Ma, L., Sha, J., Zhao, N.: Three-dimensional porous carbon materials and their composites as electrodes for electrochemical energy storage systems. Mater. Chem. Front. (2019)
32. Wang, L., Hu, X.: Recent advances in porous carbon materials for electrochemical energy storage. Chem. Asian J. (2018)
33. Li, B., Xiong, H., Xiao, Y.: Progress on synthesis and applications of porous carbon materials. Int. J. Electrochem. Sci. (2020)
34. Park, S., Ruoff, R.S.: Chemical methods for the production of graphenes. Nat. Nanotechnol. (2009)
35. Stankovich, S., Dikin, D.A., Dommett, G.H.B., Kohlhaas, K.M., Zimney, E.J., Stach, E.A., Piner, R.D., Nguyen, S.B.T., Ruoff, R.S.: Graphene-based composite materials. Nature (2006)
36. Melo, A.F.A.A., Sedenho, G.C., Osica, I., Ariga, K., Crespilho, F.N.: Electrochemical behavior of cytochrome C immobilized in a magnetically induced mesoporous framework. ChemElectroChem (2019)
37. Balasubramanian, K.: Challenges in the use of 1D nanostructures for on-chip biosensing and diagnostics: a review. Biosens. Bioelectron. (2010)
38. Decher, G.: Fuzzy nanoassemblies: toward layered polymeric multicomposites. Science **80** (1997)
39. Clarke, F.W., Irving Langmuir, B.: Constitution of solids and liquids. J. Am. Chem. Soc. (1916)
40. Langmuir, I.: Condensation and evaporation of gas molecules. Mon. Weather Rev. (1917)
41. Caseli, L.: Enzymes immobilized in Langmuir-Blodgett films: why determining the surface properties in Langmuir monolayer is important? An. Acad. Bras. Cienc. (2018)
42. Sedenho, G.C., Hassan, A., Macedo, L.J.A., Crespilho, F.N.: Stabilization of bilirubin oxidase in a biogel matrix for high-performance gas diffusion electrodes. J. Power Sources (2021)
43. Gupta, R., Chaudhury, N.K.: Entrapment of biomolecules in Sol-Gel matrix for applications in biosensors: problems and future prospects. Biosens. Bioelectron. (2007)
44. Banerjee, S., Hemraj-Benny, T., Wong, S.S.: Covalent surface chemistry of single-walled carbon nanotubes. Adv. Mater. (2005)
45. Georgakilas, V., Otyepka, M., Bourlinos, A.B., Chandra, V., Kim, N., Kemp, K.C., Hobza, P., Zboril, R., Kim, K.S.: Functionalization of graphene: covalent and non-covalent approaches, derivatives and applications. Chem. Rev. (2012)
46. Balasubramanian, K., Burghard, M.: Chemically functionalized carbon nanotubes. Small (2005)
47. Georgakilas, V., Tiwari, J.N., Kemp, K.C., Perman, J.A., Bourlinos, A.B., Kim, K.S., Zboril, R.: Noncovalent functionalization of graphene and graphene oxide for energy materials, biosensing, catalytic, and biomedical applications. Chem. Rev. (2016)
48. Blanford, C.F., Heath, R.S., Armstrong, F.A.: A stable electrode for high-potential, electrocatalytic O_2 reduction based on rational attachment of a blue copper oxidase to a graphite surface. Chem. Commun. (2007)
49. Armstrong, F.A.: Some fundamental insights into biological redox catalysis from the electrochemical characteristics of enzymes attached directly to electrodes. Electrochim. Acta (2021)
50. Hetemi, D., Noël, V., Pinson, J.: Grafting of diazonium salts on surfaces: application to biosensors. Biosensors. (2020)
51. Allongue, P., Delamar, M., Desbat, B., Fagebaume, O., Hitmi, R., Pinson, J., Savéant, J.M.: Covalent modification of carbon surfaces by aryl radicals generated from the electrochemical reduction of diazonium salts. J. Am. Chem. Soc. (1997)

52. Zollinger, H.: Reactivity and stability of arenediazonium ions. Acc. Chem. Res. (1973)
53. Sharma, R., Baik, J.H., Perera, C.J., Strano, M.S.: Anomalously large reactivity of single graphene layers and edges toward electron transfer chemistries. Nano Lett. (2010)
54. Gearba, R.I., Kim, M., Mueller, K.M., Veneman, P.A., Lee, K., Holliday, B.J., Chan, C.K., Chelikowsky, J.R., Tutuc, E., Stevenson, K.J.: Atomically resolved elucidation of the electrochemical covalent molecular grafting mechanism of single layer graphene. Adv. Mater. Interfaces (2016)
55. Zuccaro, L., Kern, K., Balasubramanian, K.: Identifying chemical functionalization on individual carbon nanotubes and graphene by local vibrational fingerprinting. ACS Nano (2015)
56. Balasubramanian, K., Zuccaro, L., Kern, K.: Tunable enhancement of Raman scattering in graphene-nanoparticle hybrids. Adv. Funct. Mater. (2014)
57. Reddy, K.S., Siva, B., Reddy, S.D., Naresh, N.R., Pratap, T.V., Rao, B.V., Hong, Y.A., Kumar, B.V., Raju, A.K., Reddy, P.M., et al.: In situ FTIR spectroscopic monitoring of the formation of the arene diazonium salts and its applications to the Heck-Matsuda reaction. Molecules (2020)
58. Harris, T.G.A.A., Heidary, N., Kozuch, J., Frielingsdorf, S., Lenz, O., Mroginski, M.A., Hildebrandt, P., Zebger, I., Fischer, A.: In situ spectroelectrochemical studies into the formation and stability of robust diazonium-derived interfaces on gold electrodes for the immobilization of an oxygen-tolerant hydrogenase. ACS Appl. Mater. Interfaces (2018)
59. Migneault, I., Dartiguenave, C., Bertrand, M.J., Waldron, K.C.: Glutaraldehyde: behavior in aqueous solution, reaction with proteins, and application to enzyme crosslinking. BioTechniques (2004)
60. Crespilho, F.N., Ghica, M.E., Gouveia-Caridade, C., Oliveira, O.N., Brett, C.M.A.: Enzyme immobilisation on electroactive nanostructured membranes (ENM): optimised architectures for biosensing. Talanta (2008)
61. Crespilho, F.N., Emilia Ghica, M., Florescu, M., Nart, F.C., Oliveira, O.N., Brett, C.M.A.: A strategy for enzyme immobilization on layer-by-layer dendrimer-gold nanoparticle electrocatalytic membrane incorporating redox mediator. Electrochem. Commun. (2006)
62. Suganya, V., Anuradha, V.: Microencapsulation and nanoencapsulation: a review. Int. J. Pharm. Clin. Res. (2017)
63. Gharsallaoui, A., Roudaut, G., Chambin, O., Voilley, A., Saurel, R.: Applications of spray-drying in microencapsulation of food ingredients: an overview. Food Res. Int. (2007)
64. Martino, C., Kim, S.H., Horsfall, L., Abbaspourrad, A., Rosser, S.J., Cooper, J., Weitz, D.A.: Protein expression, aggregation, and triggered release from polymersomes as artificial cell-like structures. Angew. Chem. Int. Ed. (2012)
65. Swi Chang, T.M.: Therapeutic applications of polymeric artificial cells. Nat. Rev. Drug Discov. (2005)
66. Li, Z., Hu, S., Cheng, K.: Chemical engineering of cell therapy for heart diseases. Acc. Chem. Res. (2019)
67. Spicer, C.D., Davis, B.G.: Selective chemical protein modification. Nat. Commun. (2014)
68. Hoyt, E.A., Cal, P.M.S.D., Oliveira, B.L., Bernardes, G.J.L.: Contemporary approaches to site-selective protein modification. Nat. Ver. Chem. (2019)
69. Morello, G., Megarity, C.F., Armstrong, F.A.: The power of electrified nanoconfinement for energising, controlling and observing long enzyme cascades. Nat. Commun. (2021)
70. Wheeldon, I., Minteer, S.D., Banta, S., Barton, S.C., Atanassov, P., Sigman, M.: Substrate channelling as an approach to cascade reactions. Nat. Chem. (2016)
71. Minton, A.P.: The influence of macromolecular crowding and macromolecular confinement on biochemical reactions in physiological media. J. Biol. Chem. (2001)
72. Zuccaro, L., Tesauro, C., Kurkina, T., Fiorani, P., Yu, H.K., Knudsen, B.R., Kern, K., Desideri, A., Balasubramanian, K.: Real-time label-free direct electronic monitoring of topoisomerase enzyme binding kinetics on graphene. ACS Nano (2015)
73. Wang, Q.H., Jin, Z., Kim, K.K., Hilmer, A.J., Paulus, G.L.C., Shih, C.J., Ham, M.H., Sanchez-Yamagishi, J.D., Watanabe, K., Taniguchi, T., et al.: Understanding and controlling the

substrate effect on graphene electron-transfer chemistry via reactivity imprint lithography. Nat. Chem. (2012)
74. Strzemińska, I., Sainte Rose Fanchine, S., Anquetin, G., Reisberg, S., Noël, V., Pham, M.C., Piro, B.: Grafting of a peptide probe for prostate-specific antigen detection using diazonium electroreduction and click chemistry. Biosens. Bioelectron. (2016)
75. Nidzworski, D., Siuzdak, K., Niedziałkowski, P., Bogdanowicz, R., Sobaszek, M., Ryl, J., Weiher, P., Sawczak, M., Wnuk, E., Goddard, W.A., et al.: A rapid-response ultrasensitive biosensor for influenza virus detection using antibody modified boron-doped diamond. Sci. Rep. (2017)
76. Schlick, T.L., Ding, Z., Kovacs, E.W., Francis, M.B.: Dual-surface modification of the tobacco mosaic virus. J. Am. Chem. Soc. (2005)
77. Paulus, G.L.C., Wang, Q.H., Strano, M.S.: Covalent electron transfer chemistry of graphene with diazonium salts. Acc. Chem. Res. (2013)
78. Balasubramanian, K., Burghard, M.: Biosensors based on carbon nanotubes. Anal. Bioanal. Chem. (2006)
79. Pumera, M.: Graphene in biosensing. Mater. Today (2011)
80. Jiang, Z., Feng, B., Xu, J., Qing, T., Zhang, P., Qing, Z.: Graphene biosensors for bacterial and viral pathogens. Biosens. Bioelectron. (2020)
81. Suvarnaphaet, P., Pechprasarn, S.: Graphene-based materials for biosensors: a review. Sensors (Switzerland) (2017)
82. Macedo, L.J.A., Iost, R.M., Hassan, A., Balasubramanian, K., Crespilho, F.N.: Bioelectronics and interfaces using monolayer graphene. ChemElectroChem (2019)
83. Frasconi, M., Mazzei, F., Ferri, T.: Protein immobilization at gold-thiol surfaces and potential for biosensing. Anal. Bioanal. Chem. (2010)
84. Ou, X., Liu, Y., Zhang, M., Hua, L., Zhan, S.: Plasmonic gold nanostructures for biosensing and bioimaging. Microchim. Acta. (2021)
85. Zhang, Y., Wang, G., Yang, L., Wang, F., Liu, A.: Recent advances in gold nanostructures based biosensing and bioimaging. Coord. Chem. Rev. (2018)
86. Lee, H., Choi, T.K., Lee, Y.B., Cho, H.R., Ghaffari, R., Wang, L., Choi, H.J., Chung, T.D., Lu, N., Hyeon, T., et al.: A graphene-based electrochemical device with thermoresponsive microneedles for diabetes monitoring and therapy. Nat. Nanotechnol. (2016)
87. Bilalis, P., Katsigiannopoulos, D., Avgeropoulos, A., Sakellariou, G.: Non-covalent functionalization of carbon nanotubes with polymers. RSC Adv. (2014)
88. Plachinda, P., Evans, D., Solanki, R.: Electrical properties of covalently functionalized graphene. AIMS Mater. Sci. (2017)
89. Balasubramanian, K., Kurkina, T., Ahmad, A., Burghard, M., Kern, K.: Tuning the functional interface of carbon nanotubes by electrochemistry: toward nanoscale chemical sensors and biosensors. J. Mater. Res. (2012)
90. Finn, M.G., Sharpless, K.B.: Click chemistry: diverse chemical function from a few good reactions - Kolb – 2001. Angew. Chem. Int. Ed. Wiley Online Library. Angew. Chem. (2001)
91. Uzman, A.: Molecular Cell Biology, 4th edn. Harvey Lodish, Arnold Berk, S. Lawrence Zipursky, Paul Matsudaira, David Baltimore and James Darnell, Freeman & Co., New York, NY, (2000), 1084 Pp., List Price $102.25, ISBN 0-7167-3136-3. Biochem. Mol. Biol. Educ. (2001)
92. Ferrari, A.C., Basko, D.M.: Raman spectroscopy as a versatile tool for studying the properties of graphene. Nat. Nanotechnol. (2013)
93. Ferrari, A.C., Meyer, J.C., Scardaci, V., Casiraghi, C., Lazzeri, M., Mauri, F., Piscanec, S., Jiang, D., Novoselov, K.S., Roth, S., et al.: Raman spectrum of graphene and graphene layers. Phys. Rev. Lett. (2006)
94. Biroju, R.K., Giri, P.K.: Defect enhanced efficient physical functionalization of graphene with gold nanoparticles probed by resonance Raman spectroscopy. J. Phys. Chem. C (2014)
95. Mak, K.F., Ju, L., Wang, F., Heinz, T.F.: Optical spectroscopy of graphene: from the far infrared to the ultraviolet. Solid State Commun. (2012)

96. Acik, M., Lee, G., Mattevi, C., Pirkle, A., Wallace, R.M., Chhowalla, M., Cho, K., Chabal, Y.: The role of oxygen during thermal reduction of graphene oxide studied by infrared absorption spectroscopy. J. Phys. Chem. C (2011)
97. Jablan, M., Buljan, H., Soljačić, M.: Plasmonics in graphene at infrared frequencies. Phys. Rev. B Condens. Matter Mater. Phys. (2009)
98. Macedo, L.J.A., Lima, F.C.D.A., Amorim, R.G., Freitas, R.O., Yadav, A., Iost, R.M., Balasubramanian, K., Crespilho, F.N.: Interplay of non-uniform charge distribution on the electrochemical modification of graphene. Nanoscale (2018)
99. Prats-Alfonso, E., Albericio, F.: Functionalization of gold surfaces: recent developments and applications. J. Mater. Sci. (2011)
100. Mallakpour, S., Soltanian, S.: Surface functionalization of carbon nanotubes: fabrication and applications. RSC Adv. (2016)
101. Zuccaro, L., Krieg, J., Desideri, A., Kern, K., Balasubramanian, K.: Tuning the isoelectric point of graphene by electrochemical functionalization. Sci. Rep. (2015)
102. Lai, S.C.S., Patel, A.N., McKelvey, K., Unwin, P.R.: Definitive evidence for fast electron transfer at pristine basal plane graphite from high-resolution electrochemical imaging. Angew. Chem. Int. Ed. (2012)
103. Zhang, G., Kirkman, P.M., Patel, A.N., Cuharuc, A.S., McKelvey, K., Unwin, P.R.: Molecular functionalization of graphite surfaces: basal plane versus step edge electrochemical activity. J. Am. Chem. Soc. (2014)
104. Güell, A.G., Cuharuc, A.S., Kim, Y.R., Zhang, G., Tan, S.Y., Ebejer, N., Unwin, P.R.: Redox-dependent spatially resolved electrochemistry at graphene and graphite step edges. ACS Nano (2015)
105. Unwin, P.R., Güell, A.G., Zhang, G.: Nanoscale electrochemistry of Sp2 carbon materials: from graphite and graphene to carbon nanotubes. Acc. Chem. Res. (2016)
106. Banks, C.E., Moore, M.R., Davies, T.J., Compton, R.G.: Investigation of modified basal plane pyrolytic graphite electrodes: definitive evidence for the electrocatalytic properties of the ends of carbon nanotubes. Chem. Commun. (2004)
107. Alsalem, H.S., Al-Goul, S.T., García-Miranda Ferrari, A., Brownson, D.A.C., Velarde, L., Koehler, S.P.K.: Imaging the reactivity and width of graphene's boundary region. Chem. Commun. (2020)
108. Banks, C.E., Compton, R.G.: Exploring the electrocatalytic sites of carbon nanotubes for NADH detection: an edge plane pyrolytic graphite electrode study. Analyst (2005)
109. Xiang, Z., Dai, Q., Chen, J.F., Dai, L.: Edge functionalization of graphene and two-dimensional covalent organic polymers for energy conversion and storage. Adv. Mater. (2016)
110. Englert, J.M., Dotzer, C., Yang, G., Schmid, M., Papp, C., Gottfried, J.M., Steinrück, H.P., Spiecker, E., Hauke, F., Hirsch, A.: Covalent bulk functionalization of graphene. Nat. Chem. (2011)
111. Yadav, A., Iost, R.M., Neubert, T.J., Baylan, S., Schmid, T., Balasubramanian, K.: Selective electrochemical functionalization of the graphene edge. Chem. Sci. (2019)
112. Iost, R.M., Crespilho, F.N., Zuccaro, L., Yu, H.K., Wodtke, A.M., Kern, K., Balasubramanian, K.: Enhancing the electrochemical and electronic performance of CVD-grown graphene by minimizing trace metal impurities. ChemElectroChem (2014)
113. Song, J., Kam, F.Y., Png, R.Q., Seah, W.L., Zhuo, J.M., Lim, G.K., Ho, P.K.H., Chua, L.L.: A general method for transferring graphene onto soft surfaces. Nat. Nanotechnol. (2013)
114. Yadav, A., Wehrhold, M., Neubert, T.J., Iost, R.M., Balasubramanian, K.: Fast electron transfer kinetics at an isolated graphene edge nanoelectrode with and without nanoparticles: implications for sensing electroactive species. ACS Appl. Nano Mater. (2020)
115. Bollella, P.: Porous gold: a new frontier for enzyme-based electrodes. Nanomaterials (2020)
116. Figueiredo, J.L., Pereira, M.F.R., Freitas, M.M.A., Órfão, J.J.M.: Modification of the surface chemistry of activated carbons. Carbon N. Y. (1999)
117. Collinson, M.M.: Nanoporous gold electrodes and their applications in analytical chemistry. ISRN Anal. Chem. (2013)

118. Ilangovan, G., Chandrasekara Pillai, K.: Mechanism of activation of glassy carbon electrodes by cathodic pretreatment. J. Solid State Electrochem. (1999)
119. Mahouche-Chergui, S., Gam-Derouich, S., Mangeney, C., Chehimi, M.M.: Aryl diazonium salts: a new class of coupling agents for bonding polymers, biomacromolecules and nanoparticles to surfaces. Chem. Soc. Rev. (2011)
120. Bélanger, D., Pinson, J.: Electrografting: a powerful method for surface modification. Chem. Soc. Rev. (2011)
121. Pellissier, M., Barrière, F., Downard, A.J., Leech, D.: Improved stability of redox enzyme layers on glassy carbon electrodes via covalent grafting. Electrochem. Commun. (2008)
122. Tamai, H., Shiraki, K., Shiono, T., Yasuda, H.: Surface functionalization of mesoporous and microporous activated carbons by immobilization of diamine. J. Colloid Interface Sci. (2006)
123. Aksoy, S., Tumturk, H., Hasirci, N.: Stability of α-amylase immobilized on poly(methyl methacrylate-acrylic acid) microspheres. J. Biotechnol. (1998)
124. Hermanson, G.T.: Bioconjugate Techniques (2008)
125. Fereiro, J.A., Porat, G., Bendikov, T., Pecht, I., Sheves, M., Cahen, D.: Protein electronics: chemical modulation of contacts control energy level alignment in gold-azurin-gold junctions. J. Am. Chem. Soc. (2018)
126. Balderston, S., Taulbee, J.J., Celaya, E., Fung, K., Jiao, A., Smith, K., Hajian, R., Gasiunas, G., Kutanovas, S., Kim, D., et al.: Discrimination of single-point mutations in unamplified genomic DNA via Cas9 immobilized on a graphene field-effect transistor. Nat. Biomed. Eng. (2021)
127. Freeley, M., Worthy, H.L., Ahmed, R., Bowen, B., Watkins, D., Macdonald, J.E., Zheng, M., Jones, D.D., Palma, M.: Site-specific one-to-one click coupling of single proteins to individual carbon nanotubes: a single-molecule approach. J. Am. Chem. Soc. (2017)

Sensitive Devices Based on Field-Effect Transistors

Henrique Antonio Mendonça Faria, Edson Giuliani Ramos Fernandes, and Nirton Cristi Silva Vieira

1 Introduction

Among the various types of transducers applied in Bioelectronics (i.e., the combination of biological materials and electronic components), Bergveld introduced the concept of field-effect transistor (FET) for neurophysiological measurements presenting to the scientific community the ion-sensitive field-effect transistor (ISFET) [1, 2]. An ISFET is a device very similar to a MOSFET (metal–oxide–semiconductor field-effect transistor). The difference between them is the non-existence of the metallic gate electrode in the ISFET, replaced by a reference electrode in a solution [3]. Figure 1 illustrates the schematic diagram of a MOSFET and an ISFET. In MOSFET, the current between the source and drain electrodes is controlled by the voltage applied to the gate. In ISFET, the current is governed by the voltage applied to the reference electrode added to the potential that appears on the metal oxide surface due to the ions accumulation at its surface [3].

Since the pioneering study of Bergveld, a remarkable advance has been made in research into developing and applying FETs as sensing devices. This chapter encompasses this progress. FETs transducers are introduced, including the ISFET and its derivatives, such as the extended gate field-effect transistor (EGFETs), the organic field-effect transistors (OFETs), nanomaterial-based field-effect transistors (NanoFETs), and as a new trend, the tunnel field-effect transistors (TFETs). Specifically, a historical overview of ISFETs, their working principle, and electrical properties are carried out. We also provide the evolution of ISFETs for one and three-dimensional nanometer devices.

E. G. R. Fernandes · N. C. S. Vieira (✉)
Institute of Science and Technology, Federal University of São Paulo, São José dos Campos, SP 12231-280, Brazil
e-mail: ncsvieira@unifesp.br

H. A. M. Faria
Institute of Chemistry, São Paulo State University (Unesp), Araraquara, SP 14800-060, Brazil

F. N. Crespilho (ed.), *Advances in Bioelectrochemistry Volume 1*,
https://doi.org/10.1007/978-3-030-94988-4_5

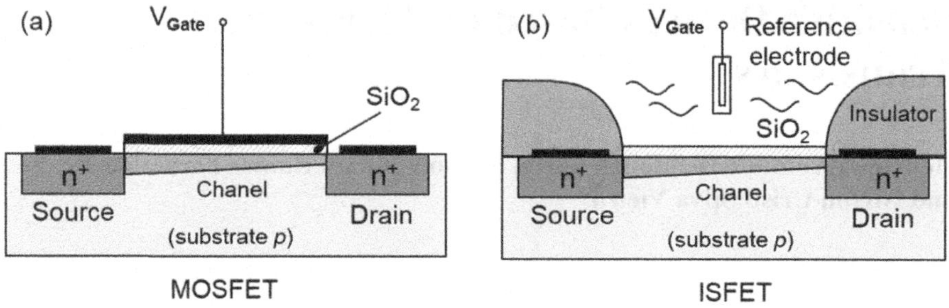

Fig. 1 **a** Schematic representation of MOSFET and **b** ISFET. Adapted from Refs. [3, 4]

2 Field-Effect Transistors Transducers

2.1 Historical Emergence of ISFET

The idea of using a FET as a microelectrode to measure the cell action potential emerged in the late 1960s [5]. This first proposal described how this device could be connected to measure the electrical potential resulting from neuronal activity. Although the application of FET allowed measuring the cell action potential, for a better understanding of the axon activity, it was necessary to quantify essential cell ions such as sodium and potassium [5]. For this, the metallic contact of the MOSFET gate had to be removed (in fact, not deposited in the microfabrication process) so that the oxide layer could directly be in contact with the biological medium. Because this device measures the concentration of ions, it was called an ion-sensitive field-effect transistor (ISFET) [1]. The signs resulting from the ionic concentrations around the muscle of the guinea-pig taenia coli were quantified in the complete study presented two years later [2]. In addition to measuring the ionic activity around the muscle, experiments were carried out to create a sensor that could measure the extracellular action potential of the flexor tibialis muscle of a locust [6]. This sensor was called OSFET, that is, without the "M" of the MOSFET. It is worth mentioning that almost at the same time, Matsuo and colleagues developed a similar device for measured pH called liquid oxide–semiconductor field-effect transistor (LOSFET) [7]. Matsuo's device differs from Bergveld's by two characteristics: the gate dielectric is composed of two layers of different materials (SiO_2 and Si_3N_4), and the data acquisition was in the saturation region of the MOSFET.

After these historical publications, many research groups started studies on ISFETs. ISFETs have been used as an alternative to measuring pH, especially under conditions in which the glass electrodes could not be applied. In the last fifty years, the use of ISFETs has added other modes of operation, including many different segments in the field of biosensors [8].

The first commercial proposal for ISFET was in the 1970s by CORDIS, a North American semiconductor company with a factory in the north of the Netherlands [8].

The device was an ISFET-based catheter containing a poly(2-hydroxyethyl methacrylate) hydrogel reference electrode. For the first time, this sensor could measure the fluctuations in blood pH resulting from breathing, a relevant parameter to control the mechanical pulmonary ventilation of patients during major surgical interventions. However, due to biocompatibility problems, the sensor was not commercialized on a large scale. Later, after the company's restructuring, the research department at CORDIS became SENTRON, which used the original idea of ISFET for several applications beyond the medical field, holding the first patent [9].

2.2 Electrical Characteristics of ISFET

The ISFET is nothing more than a MOSFET with no metallic gate, allowing the aqueous solution under analysis to contact the gate oxide directly. The operation of the MOSFET can be used to understand how ISFET works. In the MOSFET, the metallic gate and the channel region induced with negative charges form a parallel plate capacitor [4]. The SiO_2 layer works as a dielectric that enhances the capacitive effect. Between these "two plates," there will be an electric field normal to the channel. The electric field established by the gate voltage modulates the charge in the channel and defines the conductivity between the source and the drain electrodes.

The electrical behavior of the ISFET is similar to the MOSFET, with the addition of the parameters related to the oxide-solution interface. Another difference between MOSFET and ISFET is in the insulating layer. In ISFETs, additional insulating materials can be deposited on silicon dioxide to be used as a barrier to the solution while contributing to the capacitive effect [3]. Therefore, the sensor sensitivity and selectivity are determined by the choice of insulating material. The most used oxides as covering layers are SI_3N_4, Al_2O_3, SiO_2, and Ta_2O_5 [10].

Figure 2a shows a properly polarized ISFET. As in the MOSFET, there are two fundamental parameters: the drain-source current (I_{DS}) and the threshold voltage

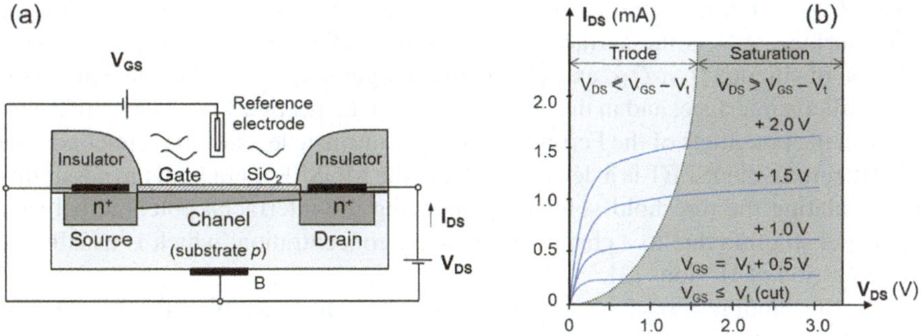

Fig. 2 **a** Schematic representation of polarized ISFET. **b** Standard I_{DS} *versus* V_{DS} curves of the regions of operation of a FET device. Adapted from Refs. [3, 4]

(V_t) [4]. The device can operate in two regimes, defined as triode and saturation regions, as shown in the I_{DS} *versus* V_{DS} (Fig. 2b).

The I_{DS} current in the triode region allows the transistor to operate as a switch and is described by Eq. (1) [4]:

$$I_{DS} = C_{ox}\frac{W}{L}\mu\left[(V_{GS} - V_t)V_{DS} - \frac{1}{2}V_{DS}^2\right] \tag{1}$$

where C_{ox} is the capacitance of the oxide per unit area, W and L are the height and length of the channel, respectively; μ is the electronic mobility in the channel, V_{GS} is the voltage between the gate and the source, V_t is the threshold voltage and V_{DS} is the voltage between drain and source. For V_t and V_{DS} constants, I_{DS} is a function of the input voltage V_{GS}. The process technology used to fabricate MOSFET/ISFET allows the parameters C_{ox}, W, L, μ, and V_t to be constant [4]. Considering the ISFET operating in the triode region, for constant and small values of V_{DS}, the quadratic term of expression (1) can be neglected, resulting in the linear relationship in which the transistor operates as a variable resistor dependent on the value of V_{GS} [4].

In the saturation region ($V_{DS} \geq V_{GS} - V_t$), it is a condition in which the transistor works as an input signal amplifier, and I_{DS} depends quadratically on V_{GS}, as indicated in Eq. (2) [4]:

$$I_{DS} = C_{ox}\frac{W}{L}\mu(V_{GS} - V_t)^2 \tag{2}$$

Other parameters relevant are related to the threshold voltage of ISFET (V_t), which is the minimum voltage required to be applied between gate and source to create an inversion of the channel depletion region, and then the channel becomes conducting. V_t is expressed as (3) [3]:

$$V_t = E_{ref} - \psi + \chi^{sol} - \frac{\Phi_{Si}}{q} - \frac{Q_{ox} + Q_{ss} + Q_B}{C_{ox}} + 2\phi_f \tag{3}$$

where E_{ref} is the potential of the reference electrode, ψ is the potential on the oxide surface, χ^{sol} is the surface dipole potential of the solvent, Φ_{Si} is the work function of silicon; Q_{ox}, Q_{ss}, and Q_B are the charges accumulated in the oxid, in the oxide-silicon interface, and in the depletion region, respectively; ϕ_f is the difference between the potentials of the Fermi level and the intrinsic level of the semiconductor [3]. Therefore, the ISFET is a device similar to the MOSFET, but with the possibility of modulating the threshold voltage by changing the interfacial potential between oxide and solution due to a change in the ionic concentration, which is usually the concentration of H^+ ions [3].

The site binding model is used to explain the sensitivity of ISFETs [11]. The surface of amphoteric oxides (A) presents species in three distinct forms: negative (AO^-), neutral (AOH), and positive (AOH_2^+) ones, according to its protonation or deprotonation. The following bindings can occur on the oxide surface:

$$AOH_2^+ \leftrightarrow AOH + H^+ \tag{4}$$

$$AOH \leftrightarrow AO^- + H^+ \tag{5}$$

In this way, the potential on the oxide surface changes according to Eq. 6:

$$\psi = \frac{2,3k_BT}{q}(pH_{Ox} - pH_{Sol}) \tag{6}$$

where k_B is the Boltzmann constant, T is the absolute temperature, q is the elementary charge; pH_{Ox} and pH_{Sol} are the pH values at the oxide surface and the analytical solution, respectively [11].

ISFET is "gated" via a reference electrode. The I_{DS} *versus* V_{DS} curve is obtained as a function of V_{GS}. Suppose $E_{ref} = 0$, curves similar to the MOSFET are obtained as a function of changes in the pH of the solution. This behavior shown in Fig. 3, indicates that the interfacial potential is related to the pH, that is, $\Psi = f$ (pH). The transfer curve for a pH-sensitive ISFET is obtained by keeping V_{DS} constant while

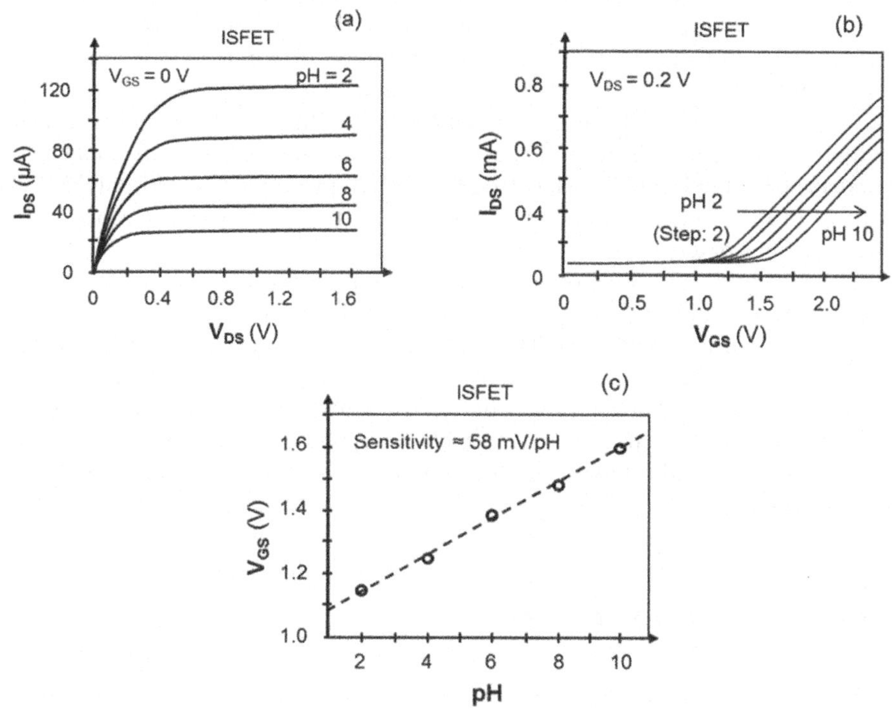

Fig. 3 **a** Expected output (I_{DS} *versus* V_{DS}), and **b** transfer (I_{DS} *versus* V_{GS}) curves of ISFET in solutions of different pH values. **c** *Vt* value in each pH value, a Nernstian behavior at a temperature of 25 °C (**c**). Adapted from Ref. [12]

V_{GS} varies in each solution with a different pH value (Fig. 3b), and the pH sensitivity is obtained due to variation of V_t. This variation depends on the gate oxide material. When it is linear with the logarithm of a particular ionic concentration, it is known as Nernstian behavior, close to 58 mV/pH at a temperature of 25 °C (Fig. 3c) [12].

The reference electrode is a necessary element in the ISFET so that the electrical circuit formed by the transistor and the electrolyte is closed. The potential of the reference electrode must remain constant regardless of what happens in the electrochemical system. Thus, potential changes will only be detected at the FET gate used as a working electrode. However, using the external reference electrode limits the miniaturization of the system and presents risks of contamination in biological or food analysis in the case of glass electrode breakdown [3]. The search for solutions came after the publication of the first works [3]. For tests, the traditional reference electrode such as Ag/AgCl offers a stable potential on larger volume samples. However, due to the difficulty in miniaturizing this reference electrode, other alternatives, such as integrating the reference electrode in the semiconductor or electrical arrangements that could provide a stable potential, started to be studied. One of the approaches was to use a differential measurement between an ISFET and another similar device that does not react with the ion to be analyzed. This device is named REFET, which an evaporated Pt pseudo reference electrode can be constructed on the substrate on which the sensors are mounted [3]. During continuous attempts to develop a stable REFET, the focus was on creating a completely inert surface. However, differential system ISFET/REFET could only succeed when both devices are electrically identical [13]. For the REFET electrically similar to the ISFET both should be coated with the same membrane. However, the REFET membrane should contain a buffer component for the ion to be analyzed. The ISFET membrane should only follow changes in the ionic concentration of the sample as in the system developed by Chudy and coworkers [14].

In summary, the operating principle of an ISFET is similar to that of a MOSFET. The voltage V_{GS} in the ISFET is applied to the reference electrode (which ideally maintains a constant potential). The potential on the metal oxide surface causes variations in the voltage V_t due to variations in the ionic concentration of the solution. Each oxide has a specific sensitivity due to its amphoteric characteristics. The so-called Nernstian materials are more suitable for applications such as ionic sensors (pH, for example) and catalytic biosensors. Other FET transducers are discussed in the following.

2.3 *EGFETs and SEGFETs*

One of the significant advantages of ISFETs is using microelectronics in their production, which enables the manufacture of biomedical devices on a large scale. On the other hand, a drawback of ISFETs is the difficulty of isolating the FET part of the reaction medium, which can influence the device's performance due to some possible ionic contamination. Another inconvenience is the high cost of developing the devices, especially when considering disposable sensors.

An elegant system for isolating the FET part of the reaction medium was proposed by Van der Spiegel and coworkers [15]. They developed the extended gate field-effect transistor (EGFET) which, comprises an outer electrode or sensitive part connected to the gate terminal of a MOSFET. Both elements, i.e., the sensitive part and the MOSFET, are fabricated on the same substrate. In this case, the microfabrication steps are still necessary for the construction of the device [15]. Alternatively, the sensitive layer can be connected to a high input impedance device, like an operational or instrumentation amplifier, both working as a voltage follower (unity-gain amplifier). According to Van der Spiegel, this strategy reduces the effects of input capacitance and leakage [15]. Later, Chi et al. improved the EGFET by connecting the sensitive layer to a commercial MOSFET. The authors showed that tin dioxide (SnO_2) has a sensitivity of ca. 56–58 mV/pH by using the commercial MOSFET CD4007UB and also proved the concept proposed by Van der Spiegel when connecting the SnO_2 membrane to the operational amplifier (LF356N) [16].

Figure 4 shows the EGFET and SEGFET configurations: a MOSFET with its gate terminal connected to an external electrode (sensitive part); both mounted on the same chip in (Fig. 4a) or with the outer electrode connected to a commercial MOSFET (Fig. 4b). The circuit is closed using a reference electrode (Ag/AgCl or similar) and an electrolyte solution. The configuration shown in Fig. 4b is often referred to in the literature as a separative-extended gate-field-effect transistor (SEGFET) because the gate is extended and separated from the FET part [17].

SEGFET configuration is more flexible for testing new materials as pH sensors. For example, Fernandes et al. showed that films formed by multilayers of poly(propylene imine) dendrimer (PPID) and nickel tetrasulfonated phthalocyanine (NiTsPC) are very stable in a pH range of 4–10. PPID/NiTsPc films onto Au-covered substrates displayed a pH sensitivity of 30 mV/pH. When deposited onto ITO (indium tin oxide), the pH sensitivity was 52.4 mV/pH, indicating that the films are porous and permeable to the H^+ ion [19].

The operating principle of an EGFET/SEGFET is similar to that described for the ISFET (or MOSFET) in the previous section, except that only the sensitive part and the reference electrode are introduced in the solution under analysis [16].

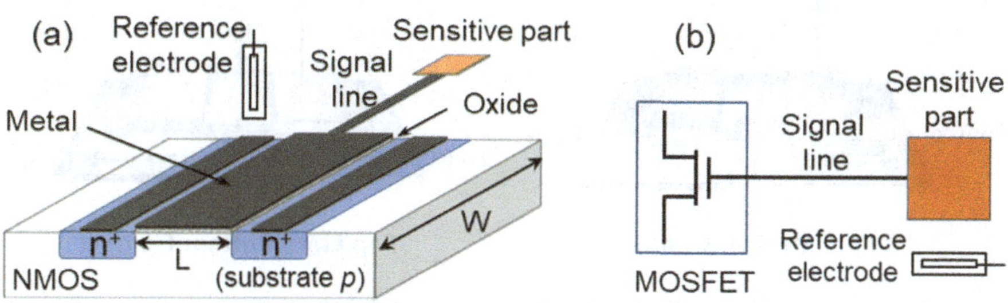

Fig. 4 **a** Configuration of an EGFET and **b** configuration of a SEGFET. Adapted from Ref. [18]

Because only the sensitive part needs to be built, the EGFET/SEGFET configuration has some advantages over traditional ISFET: (i) the measurement setup is simplified, (ii) new materials can be easily tested as pH sensors (and other ions), (iii) the possible gate ionic contamination can be avoided, and (iv) the immobilization of biomolecules is facilitated through the functionalization of just sensitive part, allowing the development of new biosensors [19].

2.4 OFETs

The organic field-effect transistors (OFETs) were proposed in the 1980s [20]. In OFETs, the active layer is an organic semiconductor or both; the active and dielectric layers are organic materials (ordinarily organic thin films) [21]. Figure 5 shows the four typical configurations of OFETs due to the conducting channel position: along with the insulating/semiconductor interface (Bottom-gate/Bottom-contact and Top-gate/Top-contact), at the base of the semiconductor (Top-gate/Bottom-contact) or the top of the semiconductor (Bottom-gate/Top-contact) [22].

The first ion-sensitive organic-based field-effect transistor (ISOFET) for monitoring pH was reported by Bartic and colleagues [23]. In this case, an organic semiconductor (poly(3-hexylthiophene)) was used as a semiconductor layer and silicon nitride as a dielectric layer (a silicon wafer was used as a support for the transistor). Loi et al. reported the use of a plastic 900-nm-thick thin foil (Mylar™ from Dupont)

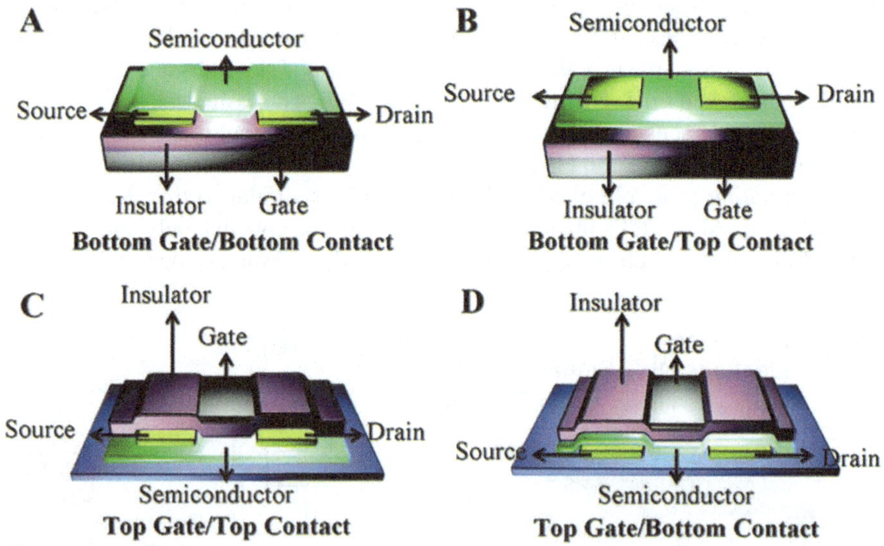

Fig. 5 Four types of OFETs. **a** Bottom gate/bottom contact, **b** bottom gate/top contact, **c** top gate top contact, and **d** top gate bottom contact. Reprinted with permission from Ref. [22]. Copyright (2012) American Chemical Society

as substrate and gate insulator (dielectric rigidity of 10^5 V/cm) [24]. Thus, Pentacene 50-nm-thick film was used as a p-channel of the FET working in accumulation mode in totally ISOFET. As well as inorganic FETs, the device sensitivity is due to the charge density in the FET conducting channel, between source and drain electrodes, by controlling the gate voltage [24]. Besides, the transistor performance is strongly dependent on the solvent used in the organic film deposition. The main problem in the organic devices is the operating voltage in solution, which can promote water electrolysis, which can be overcome by an electrolyte-gated organic field-effect transistor (EGOFET) configuration, similar to ISFET configuration with an organic layer as sensing gate. In the case, with the use of a dielectric with higher capacitance (for example, Ta_2O_5 instead Si_3N_4) and low operating voltage (<1 V), due to the electrical double layer capacitance of a few tens of $\mu F/cm^2$ [24].

Some strategies are commonly used to avoid the problem mentioned above: (i) enhance the current density in the OFET channel, enhancing carrier accumulation over the insulating material. The semiconductor channel must allow the transport of both ions and electrical charges (negative charges or electrons in n-type FET and positive charges or holes in p-type FET). Electrochemical reactions can induce positive or negative charges in the gate, changing the concentration in the conducting channel. (ii) Enhance the capacitance of the dielectric layer, assured by the use of high-k dielectric materials or using a dielectric thin-film. The semiconducting layer is responsible for carrier accumulation and depends on the capacitance of the dielectric. The induced charge can be controlled by the thickness and relative permittivity of the dielectric layer. (iii) Performing detection of biological events in a "dry" device. For example, an OFET-based genosensor was developed by immobilizing single-stranded DNA molecules (ssDNA) on the organic layer. After hybridizing the oligonucleotides in a solution containing the complementary DNA of interest, the device was dry. The measurement of the shift in transistor saturation current was used for DNA hybridization detection [25]. Another problem is the stability of the aqueous potential for charge transport through the channel, once hydration is essential for ion transport. Someya and coworkers protected the source and drain electrodes from contact with water [26]. Using a more compact layer and low density of defects by ordinated film, such as Langmuir–Blodgett film, can also solve the problem [27].

In summary, the advantages in the use of OFETs comparatively to ISFETs or EGFETs are based on their: high sensitivity, large-scale processability, relatively low cost, flexibility, possible biocompatibility, printability, the potential for wearable devices, and the possibility of exploring the device surface functionalization.

2.5 NanoFETs

The study and manipulation of matter on a nanometric scale (i.e., Nanotechnology) paved the way for a new generation of FETs. Nanomaterial-based field-effect transistors (NanoFETs), that is, FETs that use a nanomaterial as the transistor channel,

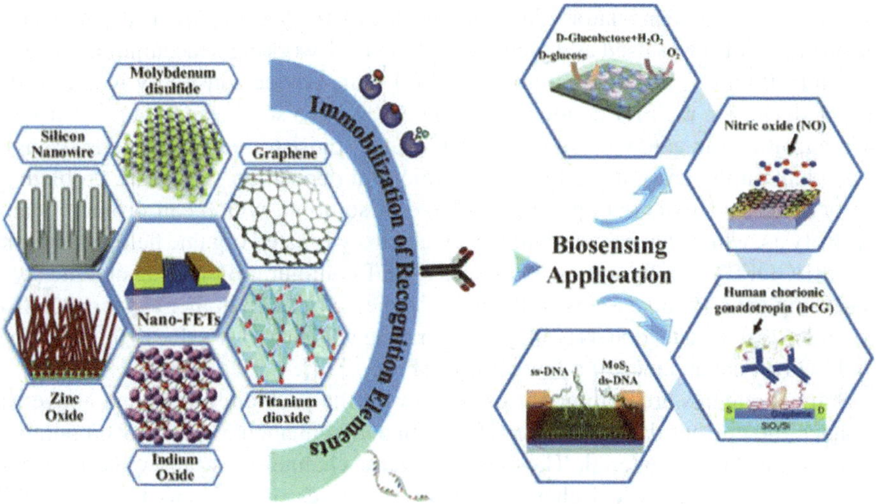

Fig. 6 Schematic representation of NanoFETs and their application as biosensors. The transistor channel is a nanomaterial. Reprinted with permission from Ref. [30]. Copyright (2019) Springer Nature

have different or improved physical–chemical properties than conventional ones due to their increased surface-to-volume ratio [28].

In principle, different types of nanostructures can be used to manufacture NanoFETs, including one-dimensional (1D) nanotubes, nanowires (NWs), and nanorods, and more recently, two-dimensional (2D) nanosheets [29]. It is noteworthy that here we do not consider ISFETs modified or decorated with nanomaterials as NanoFETs. In other words, here we consider when nanomaterials are the channel component of the FET, as displayed schematically in Fig. 6 [30]. We also include here FETs formed by the organic nanomaterials, mainly graphene and carbon nanotubes.

In NanoFETs, the nanosized channel is directly exposed to the measurement solution, making these devices much more sensitive than traditional ISFETs. The device can be gated by a bottom gate dielectric or via a reference electrode in a solution-gated configuration. In the latter, the electrical double layer (EDL) acts as the dielectric, and the device operates at lower gate voltage because almost voltage applied to the reference electrode drops in the small EDLs that form at the reference electrode/solution and solution/nanomaterial interfaces [31].

In nanomaterials, their electrical properties vary in relation to their macroscale counterpart which can be explained by variation in the Density of States (σ) of the material, assuming that σ is proportional to the energy (E) and dimensionality (D) as [32]:

$$\sigma \propto \sqrt{E}^{D-2} \tag{7}$$

Furthermore, a change in the material's bandwidth (bandgap) is expected at a low scale.

The interest in using NanoFETs has attracted significant attention due to the discovery of new sensing mechanisms, possible integration of the detection part with conventional semiconductor techniques, high spatial resolution, high conductivity, high chemical stability, and possible integration with high-k dielectric materials [30].

2.5.1 1D NanoFETs

The most 1D nanostructure used in NanoFETs are nanowires, including a range of semiconductors from group IV, group III–V, group II–VI, and metal oxide semiconductors. Organic nanowires, like carbon nanotubes and conductive polymer nanowires, are also utilized. But the silicon industry's well-established manufacturing and processing techniques make silicon nanowires (SiNWs) one of the most widely used nanomaterials in NanoFETs [33].

There are two main fabrication techniques for preparing SiNWs, top-down and bottom-up. The first process consists of the patterning of the SiNW from monocrystalline silicon via photolithography and electron-beam processes. The second method starts with the growth of SiNWs by chemical vapor deposition (CVD), followed by the deposition of SiNW in a SiO_2/Si substrate and fabrication of the devices by photo or electron beam lithography methods. Each methodology has its advantages and disadvantages in terms of cost, time, and reproducibility of NanoFETs [33].

SiNWs are also representative examples of NanoFETs because their physical–chemical properties have been widely explored and controlled, as demonstrated over the years by Prof. Lieber's pioneering group [34]. The working principle of a NanoFET is very similar to a conventional FET. If no gate voltage is applied, there is no appreciable current in the nanochannel. When the gate voltage is applied and exceeds the threshold voltage, charge carriers are induced in the nanomaterial (electrons for an n-type nanowire or holes for a p-type). Thus, the source-drain current can then be turned on and off by gate voltage. For example, boron-doped SiNWs-FETs, when functionalized with 3-aminopropyltriethoxysilane, exhibited a pH-dependent conductance due to protonation and deprotonation on their surface. A potential change modulates the nanochannel conductance on their surface as a result of a chemical reaction [33].

In particular, carbon 1D structure has low heat dissipation and results in FET devices with low operating temperature with a high current response at low voltages [35]. Despite the broad applicability of inorganic nanowires in FETs, carbon nanotubes (CNTs) represent the starting point for the realization of NanoFETs [36]. The surface single-molecule sensitivity of the CNTs with label-free detection and fast response speed has excellent potential for real-time sensors [35]. The interactions with a single molecule are possible because of the highly sensitivity of carbon material to changes in their chemical environment: the electrical properties of carbon nanotubes, such as electrical conductance, depends on the pH, the ionic strength, and ions present in electrolyte voltages [35].

Several methods can obtain CNTs, but the most common reported techniques are: arc discharge, chemical vapor deposition (CVD), and laser ablation [37]. Depending on the process, is possible to synthesize single-walled carbon nanotubes (SWCNTs), multi-walled carbon nanotubes (MWCNTs), or a mixture of both, which will be composing the channel of the FET. Furthermore, it is possible to obtain CNTs with metallic or semiconductor electrical properties [37].

For applications in FET, semiconductor SWCNs are more adequate due to the control of channel properties in terms of the semiconductor bandgap and, then, electrical properties. However, in large-scale synthesis, both types of CNTs, metallic and semiconductor, are obtained, and it is so hard to synthesized pure samples of SWCNT semiconductors [37]. Then, a disadvantage of CNTs-FETs is their low reproducibility which requires high control of the nanotubes overall electrical properties (metallic or semiconductor) and their dimensions. Type-p semiconductor SWCNT is more common in FET-based devices. The adsorption of molecules on the CNT surface can introduce defect levels near the conduction band and modulate the charge transport in the nanotube [38]. Then, the direct detection of charge transfer in the nanotube surface (or doping process) by adsorbed molecules makes the CNTs a transducer component in FET for label-free and real-time detection, enabling its application in studies of molecular recognition processes.

2.5.2 2D NanoFETs

As described in the previous section, 1D nanostructures are highly sensitive to charge changes on their surface. However, the high cost of fabrication, reproducibility, and low integrability, especially for a good alignment of nanotubes, has been a problem for the practical realization (or mass fabrication) of these devices. In the opposite direction, 2D nanomaterials have shown promise for the fabrication of NanoFETs because they are also highly sensitive to changes in the electric field in their surroundings. Furthermore, because they are flat and easy to process, they allow the fabrication of highly integrated FETs on a large scale and at a relatively low cost, increasing interest in these NanoFETs for biomedical applications, especially biosensors.

In 2004, Novoselov and coworkers demonstrated the isolation of graphene and its application as FET [39]. Graphene comprises an atomic monolayer of sp^2 carbons arranged in a 2D hexagonal crystal lattice and is the precursor of the other forms of carbon [40]. Graphene is mainly obtained from three different routes: mechanical exfoliation, chemical vapor deposition (CVD), and chemical exfoliation. As proposed by Novoselov, mechanical exfoliation produces the highest quality graphene, but with low yield and reproducibility [39]. Although producing graphene of inferior quality to mechanical exfoliation, the CVD technique is more suitable for large-scale production, being compatible with microfabrication processes in a clean room [31]. Chemical exfoliation is used to obtain graphene oxide (GO) in large quantities and at a low cost from the chemical exfoliation of graphite [41]. GO is an insulating material, but its reduction produces a graphene derivative with high electrical conductivity. Although reduced graphene oxide (rGO) has lower electrical properties than pristine graphene, it has been widely used in FETs due to its low cost and the presence of functional groups that can be used to immobilize molecules [42].

The structure of the device is similar to that of the general NanoFET shown in Fig. 6, i.e., the graphene or rGO is deposited on the two metallic source and drain electrodes. However, the *I versus V* (Fig. 7) curves of graphene FET (gFET) are different from traditional semiconductors. As graphene is a zero-bandgap semimetal, in the output curve, there is no saturation region and, the transfer curve (I_{DS} *versus* V_{GS} at constant V_{DS}) exhibits an ambipolar conduction behavior around a charge-neutrality point (CNP), also known as the Dirac point [31, 39]. In other words, the lack of a bandgap makes a gFET impossible to turn off; depending on the voltage applied to the gate, there is conduction by positive or by negative charges and a point where the density of the two carriers is equal [31, 39]. This fact compromises the switching applications of gFETs. On the other hand, they are widely used in bioelectronics for presenting carrier mobility far superior to conventional semiconductors [43].

The significant progress in graphene motivated the application of other 2D materials in FETs. 2D transition metal dichalcogenides (TMDs) are widely used in NanoFETs. They are monolayers of atomically thin semiconductors type MX_2, where M is a transition metal and X is a chalcogen [44]. The methods for synthesizing TMDs are similar to those for graphene. The unique characteristics of TMDs like a direct bandgap, atomic thickness, excellent mechanical and electrical properties make them ideal for biomedical applications [44].

TMDs are almost as thin, transparent, and flexible as graphene. Many of them are semiconductors by nature with great potential to applications in FETs expected to be more efficient than conventional silicon-based ones. Unlike graphene, the presence of a bandgap in TMDs enables high switching ratios. This property makes 2D TMDs new candidates for the next-generation bioelectronics [45].

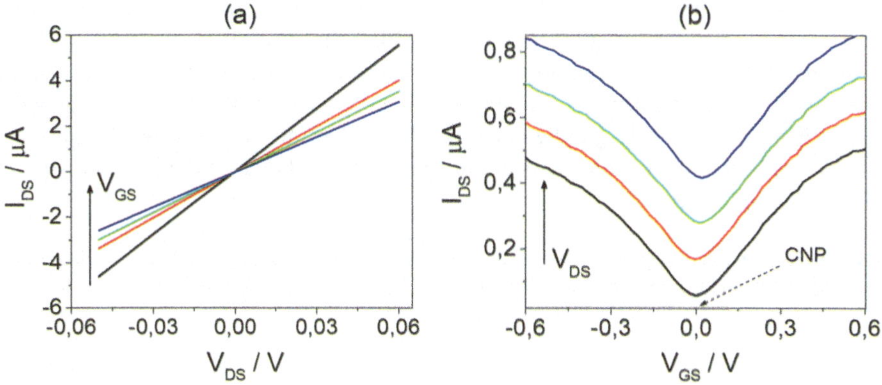

Fig. 7 Idealized *I versus V* curves of graphene FET. **a** Output curves and **b** transfer curves. Adapted from Ref. [31]

2.6 New Trends: Tunable FETs

In contrast to MOSFETs, where the current switching process involves thermionic injection, the injection mechanism in a tunnel field-effect transistor (TFET) is inter-band tunneling, through which charge carriers are transferred from one energy band to another at a strongly doped $p^{+}-n^{+}$ junction [46]. Due to this tunneling mechanism, TFETs maintain excellent switching characteristics even at high temperatures. These devices can achieve a 100-fold power reduction compared to conventional transistors. Because TFETs have low-off currents, they are ideal for low power systems [46]. TFETs represent a new trend for biomedical applications [47]. Among the types of TFET, the nanowire tunnel field-effect transistor (NW-TFET) has great potential for applications in biosensors, since they have ultra-low operating power [48]. Another advantage of NW-TFET is the possibility of construction with multi-gates. Dual-port NW-TFET devices improve operations at very high frequencies. As biosensors require high sensitivity to changes in biomolecular bonds, this NW-TFET differentiated structure results in the best performance regarding the subthreshold slope [48].

3 Final Remarks

Bergveld's proposal led to the development of a series of field-effect-based devices. Whether to measure ion activity in chemical environments or for application as biosensors. Among these devices stand out, the ISFET with different ion-sensitive materials, its derivation EGFET/SEGFET mainly aims to isolate the FET part of the chemical environment and the OFETs. These last ones especially provide a wide range of processability, low cost, flexibility, printability, the potential for wearable devices, among others. Nanotechnology enabled the development of NanoFETs, primarily with 1D nanostructures such as carbon nanotubes and semiconductor nanowires. More recently, 2D nanomaterials were used to develop a new generation of NanoFETs, where graphene and its derivatives are the most used materials. Finally, each FET device has its peculiarities, and the choice of device will depend on the final application.

Acknowledgements The authors are grateful for the financial support provided by São Paulo Research Foundation (FAPESP: 18/07508-3).

References

1. Bergveld, P.: Development of an ion-sensitive solid-state device for neurophysiological measurements. IEEE Trans. Biomed. Eng. 70–71 (1970)
2. Bergveld, P.: Development, operation, and application of the ion-sensitive field-effect transistor as a tool for electrophysiology. IEEE Trans. Biomed. Eng. 342–351 (1972)
3. Bergveld, P.: Thirty years of ISFETOLOGY: what happened in the past 30 years and what may happen in the next 30 years. Sens. Actuators, B Chem. **88**, 1–20 (2003). https://doi.org/10.1016/S0925-4005(02)00301-5
4. Sedra, A.S., Sedra, D.E.A.S., Smith, K.C.: Microelectronic Circuits. Oxford University Press, New York (1998)
5. Bergveld, P.: New amplification method for depth recording. IEEE Trans. Biomed. Eng. 102–105 (1968)
6. Bergveld, P., Wiersma, J., Meertens, H.: Extracellular potential recordings by means of a field effect transistor without gate metal, called OSFET. IEEE Trans. Biomed. Eng. 136–144 (1976)
7. Matsuo, T., Wise, K.D.: An integrated field-effect electrode for biopotential recording. IEEE Trans. Biomed. Eng. 485–487 (1974)
8. Toumazou, C., Georgiou, P.: Piet Bergveld-40 years of ISFET technology: from neuronal sensing to DNA sequencing. Electron. Lett. **47**, S7–S12 (2011)
9. Ligtenberg, H.C.G., Leuveld, J.G.M.: ISFET-based measuring device and method for correcting drift (1987)
10. Van Hal, R.E.G., Eijkel, J.C.T., Bergveld, P.: A novel description of ISFET sensitivity with the buffer capacity and double-layer capacitance as key parameters. Sens. Actuators B Chem. **24**, 201–205 (1995)
11. Yates, D.E., Levine, S., Healy, T.W.: Site-binding model of the electrical double layer at the oxide/water interface. J. Chem. Soc. Faraday Trans. 1 Phys. Chem. Condens. Phases. **70**, 1807–1818 (1974)
12. Chin, Y.-L., Chou, J.-C., Sun, T.-P., Liao, H.-K., Chung, W.-Y., Hsiung, S.-K.: A novel SnO_2/Al discrete gate ISFET pH sensor with CMOS standard process. Sens. actuators B Chem. **75**, 36–42 (2001)
13. Matsuo, T., Nakajima, H.: Characteristics of reference electrodes using a polymer gate ISFET. Sens. Actuators. **5**, 293–305 (1984)
14. Chudy, M., Wróblewski, W., Brzózka, Z.: Towards REFET. Sens. Actuators B Chem. **57**, 47–50 (1999)
15. Van der Spiegel, J., Lauks, I., Chan, P., Babic, D.: The extended gate chemically sensitive field effect transistor as multi-species microprobe. Sens. Actuators **4**, 291–298 (1983). https://doi.org/10.1016/0250-6874(83)85035-5
16. Chi, L.L., Chou, J.C., Chung, W.Y., Sun, T.P., Hsiung, S.K.: Study on extended gate field effect transistor with tin oxide sensing membrane. Mater. Chem. Phys. **63**, 19–23 (2000). https://doi.org/10.1016/S0254-0584(99)00184-4
17. Vieira, N.C.S., Fernandes, E.G.R., Queiroz, A.A.A.D., Guimarães, F.E.G., Zucolotto, V.: Indium tin oxide synthesized by a low cost route as SEGFET pH Sensor. Mater. Res. **16** (2013). https://doi.org/10.1590/S1516-14392013005000101
18. Fernandes, E.G.R., Vieira, N.C.S., De Queiroz, A.A.A., Guimarães, F.E.G., Zucolotto, V.: Immobilization of poly(propylene imine) dendrimer/nickel phtalocyanine as nanostructured multilayer films to be used as gate membranes for SEGFET pH sensors. J. Phys. Chem. C. **114**, 6478–6483 (2010). https://doi.org/10.1021/jp9106052
19. Pullano, S.A., Critello, C.D., Mahbub, I., Tasneem, N.T., Shamsir, S., Islam, S.K., Greco, M., Fiorillo, A.S.: EGFET-based sensors for bioanalytical applications: a review (2018)
20. Ebisawa, F., Kurokawa, T., Nara, S.: Electrical properties of polyacetylene/polysiloxane interface. J. Appl. Phys. **54**, 3255–3259 (1983)
21. Kergoat, L., Piro, B., Berggren, M., Horowitz, G., Pham, M.-C.: Advances in organic transistor-based biosensors: from organic electrochemical transistors to electrolyte-gated organic field-effect transistors. Anal. Bioanal. Chem. **402**, 1813–1826 (2012)

22. Wang, C., Dong, H., Hu, W., Liu, Y., Zhu, D.: Semiconducting π-conjugated systems in field-effect transistors: a material odyssey of organic electronics. Chem. Rev. **112**, 2208–2267 (2012)
23. Bartic, C., Palan, B., Campitelli, A., Borghs, G.: Monitoring pH with organic-based field-effect transistors. Sens. Actuators B Chem. **83**, 115–122 (2002)
24. Loi, A., Manunza, I., Bonfiglio, A.: Flexible, organic, ion-sensitive field-effect transistor. Appl. Phys. Lett. **86**, 103512 (2005)
25. Zhang, Q., Subramanian, V.: DNA hybridization detection with organic thin film transistors: toward fast and disposable DNA microarray chips. Biosens. Bioelectron. **22**, 3182–3187 (2007)
26. Someya, T., Dodabalapur, A., Gelperin, A., Katz, H.E., Bao, Z.: Integration and response of organic electronics with aqueous microfluidics. Langmuir **18**, 5299–5302 (2002)
27. Sizov, A.S., Agina, E.V., Gholamrezaie, F., Bruevich, V.V., Borshchev, O.V., Paraschuk, D.Y., de Leeuw, D.M., Ponomarenko, S.A.: Oligothiophene-based monolayer field-effect transistors prepared by Langmuir-Blodgett technique. Appl. Phys. Lett. **103**, 134_1 (2013)
28. Nehra, A., Singh, K.P.: Current trends in nanomaterial embedded field effect transistor-based biosensor. Biosens. Bioelectron. **74**, 731–743 (2015)
29. Mao, S., Chang, J., Pu, H., Lu, G., He, Q., Zhang, H., Chen, J.: Two-dimensional nanomaterial-based field-effect transistors for chemical and biological sensing. Chem. Soc. Rev. **46**, 6872–6904 (2017)
30. Syedmoradi, L., Ahmadi, A., Norton, M.L., Omidfar, K.: A review on nanomaterial-based field effect transistor technology for biomarker detection. Microchim. Acta. **186**, 1–23 (2019)
31. Vieira, N.C.S., Borme, J., MacHado, G., Cerqueira, F., Freitas, P.P., Zucolotto, V., Peres, N.M.R., Alpuim, P.: Graphene field-effect transistor array with integrated electrolytic gates scaled to 200 mm. J. Phys. Condens. Matter. **28** (2016). https://doi.org/10.1088/0953-8984/28/8/085302
32. Dick, R.: Dimensional effects on densities of states and interactions in nanostructures. Nanoscale Res. Lett. **5**, 1546–1554 (2010)
33. Chen, K.-I., Li, B.-R., Chen, Y.-T.: Silicon nanowire field-effect transistor-based biosensors for biomedical diagnosis and cellular recording investigation. Nano Today **6**, 131–154 (2011)
34. Cui, Y., Wei, Q., Park, H., Lieber, C.M.: Nanowire nanosensors for highly sensitive and selective detection of biological and chemical species. Science **80**(293), 1289–1292 (2001). https://doi.org/10.1126/science.1062711
35. Yao, X., Zhang, Y., Jin, W., Hu, Y., Cui, Y.: Carbon nanotube field-effect transistor-based chemical and biological sensors. Sensors **21**, 995 (2021)
36. Alvi, P.A., Lal, K.M., Siddiqui, M.J., Naqvi, A.H.: Carbon nanotubes field effect transistors: a review (2005)
37. Sinnott, S.B., Andrews, R.: Carbon nanotubes: synthesis, properties, and applications. Crit. Rev. Solid State Mater. Sci. **26**, 145–249 (2001)
38. Qian, Q., Li, G., Jin, Y., Liu, J., Zou, Y., Jiang, K., Fan, S., Li, Q.: Trap-state-dominated suppression of electron conduction in carbon nanotube thin-film transistors. ACS Nano **8**, 9597–9605 (2014)
39. Novoselov, K.S., Geim, A.K., Morozov, S.V., Jiang, D., Zhang, Y., Dubonos, S.V., Grigorieva, I.V., Firsov, A.A.: Electric field effect in atomically thin carbon films. Science **80** 306, 666–669 (2004)
40. Novoselov, K.S., Fal'Ko, V.I., Colombo, L., Gellert, P.R., Schwab, M.G., Kim, K.: A roadmap for graphene. Nature **490**, 192–200 (2012). https://doi.org/10.1038/nature11458
41. Dreyer, D.R., Park, S., Bielawski, C.W., Ruoff, R.S.: The chemistry of graphene oxide. Chem. Soc. Rev. **39**, 228–240 (2010)
42. dos Santos, F.A., Vieira, N.C.S., Zambianco, N.A., Janegitz, B.C., Zucolotto, V.: The layer-by-layer assembly of reduced graphene oxide films and their application as solution-gated field-effect transistors. Appl. Surf. Sci. 148698 (2020)
43. Hess, L.H., Seifert, M., Garrido, J.A.: Graphene transistors for bioelectronics. Proc. IEEE **101**, 1780–1792 (2013)
44. Manzeli, S., Ovchinnikov, D., Pasquier, D., Yazyev, O.V., Kis, A.: 2D transition metal dichalcogenides. Nat. Rev. Mater. **2**, 1–15 (2017)

45. Hu, H., Zavabeti, A., Quan, H., Zhu, W., Wei, H., Chen, D., Ou, J.Z.: Recent advances in two-dimensional transition metal dichalcogenides for biological sensing. Biosens. Bioelectron. **142**, 111573 (2019)
46. Ionescu, A.M., Riel, H.: Tunnel field-effect transistors as energy-efficient electronic switches. Nature **479**, 329–337 (2011)
47. Reddy, N.N., Panda, D.K.: A comprehensive review on tunnel field-effect transistor (TFET) based biosensors: recent advances and future prospects on device structure and sensitivity. Silicon 1–16 (2020)
48. Gao, A., Lu, N., Wang, Y., Li, T.: Robust ultrasensitive tunneling-FET biosensor for point-of-care diagnostics. Sci. Rep. **6**, 1–9 (2016)

Supramolecular Electrochemistry: Recent Trends and Perspectives

Aurileide Maria Bispo Frazão Soares, Caio Lenon Chaves Carvalho, Gleison de Andrade Rodrigues, Roberto A. S. Luz, Everson T. S. Gerôncio, and Welter Cantanhêde

1 General Overview

This chapter describes recent advances in the construction of functional self-assembled nanomaterials from molecular building blocks using supramolecular chemistry concepts and suitable self-assembly processes. Supramolecular self-assembly is the spontaneous arrangement of molecular components into well-defined nanostructures that can be stabilized by general interaction processes: (i) dynamic covalent (e.g. imine and disulfide bonding), (ii) reversible noncovalent interactions (e.g. electrostatic interactions, hydrogen bonds, hydrophobic, and hydrophilic interactions, and van der Waals interactions), and (iii) mutual associative approach [1].

The interaction of components at the molecular level allows the formation of self-assembled nanomaterials with diversified structures and functionalities. These nanoarchitectures can be achieved through molecular associative recognition, self-organization, and dynamic nature [2]. For this proposal, the use of molecular building blocks performs a crucial role in supramolecular self-assembly process due to their versatility and functionality.

Building blocks can also be oriented towards the development of robust architectures onto electrode, such as self-assembled monolayers (SAMs), layer-by-layer self-assembly, and electrochemical deposition. In this case, the electrochemical supramolecular approach can be utilized to investigate kinetic and thermodynamic study of electrochemical redox and electron transfer process from self-assembled organized structures as well as investigating the electron transport mechanism. We

A. M. B. F. Soares · C. L. C. Carvalho · G. de Andrade Rodrigues · R. A. S. Luz · E. T. S. Gerôncio · W. Cantanhêde (✉)
Department of Chemistry, Federal University of Piauí, Teresina, PI 64049-550, Brazil
e-mail: welter@ufpi.edu.br

F. N. Crespilho (ed.), *Advances in Bioelectrochemistry Volume 1*,
https://doi.org/10.1007/978-3-030-94988-4_6

will describe how electrochemistry supramolecular can be utilized to control electrochemical conversion, affect the structure, and supramolecular systems properties. Special emphasis will be given to the electrochemistry mechanism of host–guest complexes, substrates modified, monolayer, and multilayer assemblies. Finally, we conclude with an outlook toward future developments of self-assembled nanomaterials for constructions of new (bio)sensor, energy storage, electronic, and electrochemical devices.

2 Versatile Building Blocks Used for Supramolecular Self-Assembly

Different individual components can be used for construction of supramolecular nanostructures, such as metallic nanoparticles (MNPs) [3], coordination compounds [4, 5], semiconductive oxides [6, 7], carbon-based materials [8], and biomaterials [9]. Figure 1 illustrates the shapes and structures of some building blocks discussed in this chapter. The following subsections have been organized in order to show the importance of the components in the context of supramolecular chemistry.

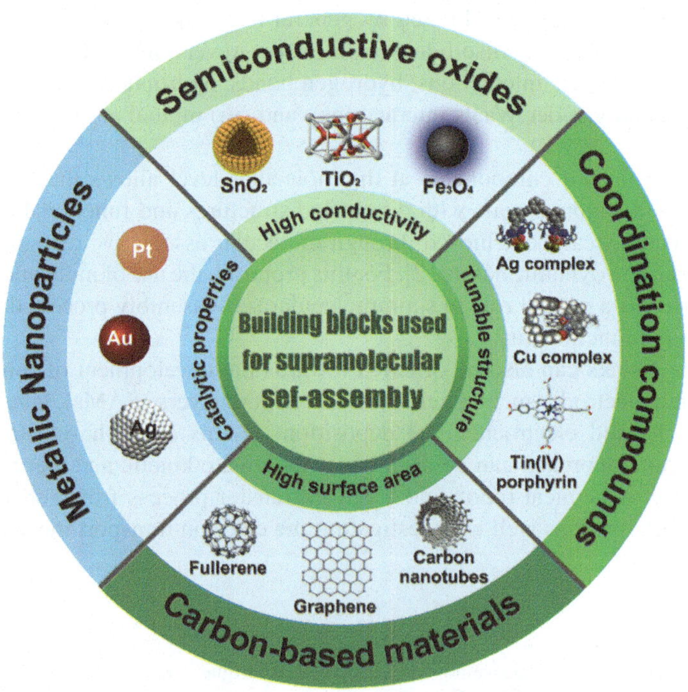

Fig. 1 Scheme of the versatile building blocks used in supramolecular chemistry. Adapted and reprinted with permission from Ref. [10, 11]. Copyright 2014 Royal Society of Chemistry

2.1 Metallic Nanoparticles

Metallic nanoparticles (MNPs) have received considerable attention of scientific community owing to their catalytic properties, antibacterial activity [12]; high surface-to-volume ratio and unique optical and electrical properties [13, 14]. The surface functionalization of MNPs enables the use of these as molecular building blocks for supramolecular chemistry.

Among the most used MNPs by literature we highlight the gold nanoparticles (Au NPs) due to their low toxicity, high chemical stability, and their unique and tunable plasmonic properties [13]. Zhang [3] used Au NPs as individual components to develop an improved glucose sensor. The high catalytic activity exhibited by Au NPs improved the charge transfer between the glucose and the GOX enzyme during the oxidation–reduction reaction resulting in a doubling of electrode sensitivity. The electrode containing the Au NPs showed high sensitivity (216.9 μA m mol L^{-1}) and low detection limit (1×10^{-6} mol L^{-1}), confirming that the electrode is a potential candidate for the non-invasive monitoring of glucose [3]. Golub et al. [15] used Pt NPs, CdS NPs, and Au NPs functionalized with an anticocaine aptamer subunit in the detection electrochemical, photoelectrochemical, and surface plasmon resonance (SPR) detection of the cocaine, respectively. The formation of the supramolecular complex obtained by the interaction between the Pt NPs-aptamer and the cocaine subunits allowed the detection of cocaine by the electro catalyzed reduction of H_2O_2. The photoelectrochemical detection of cocaine was achieved due to the photocurrents generated by the complex formed between the cocaine subunits and the CdS-NPs-aptamer, in the presence of amine triethanol as a hole-eliminator. The supramolecular Au-NPs-aptamer subunits-cocaine complex generated on the Au support allowed the SPR detection of cocaine through the reflectance changes when stimulated by the electronic coupling between the localized plasmon of the Au-NPs and the surface plasmon wave. All the developed aptasensors allowed an analysis of cocaine with a detection limit in the range of 10^{-6}–10^{-5} mol L^{-1} [15].

Lopes et al. [16] developed gold (Au NPs) and silver (Ag NPs) nanoparticles through a green route, using tannic acid (TA) as a reducing and stabilizing agent. The supramolecular interactions between Au NPs, Ag NPs and TA were responsible for the bathochromic shift (red-shifted) of the surface plasmonic resonance (SPR) bands of these nanoparticles [16].

2.2 Semiconductor Oxides

Metal oxide semiconductors represent a diverse and fascinating class of materials because of their high surface volume ratio, abundant active sites, and surface defects, which provide good detection limits, making them attractive for the development of materials with applications in various areas [17].

The magnetite nanoparticles (Fe_3O_4 NPs) have shown good biocompatibility, low toxicity, magnetic properties, easy to obtain, high surface area, and suitable functionalization, making them attractive for various chemical and bio related applications, such as enzyme interaction, drug delivery, wastewater treatment, and toxic analytes sensing [18, 19]. For example, Lima and collaborators [20] used Fe_3O_4 NPs, poly(allylamine hydrochloride) (PAH), poly(vinyl sulfonic acid) (PVS) and polypyrrole (Ppy) to develop layer-by-layer (LbL) films deposited onto gold substrate with three distinct supramolecular architectures: $(PAH/Ppy)_n$, (Fe_3O_4 $NPs/PVS)_n$, and (Fe_3O_4 $NPs/Ppy)_n$. The (Fe_3O_4 $NPs/Ppy)_n$ film showed stronger electrostatic interaction, higher conductivity, as well as a more homogeneous substrate covering in comparison to the other developed films, demonstrating a synergistic effect between Fe_3O_4 NPs and ppy. The electrochemical and surface plasmon resonance analysis confirming that Fe_3O_4 NPs-Ppy film can be used as an electrochemical or optical non-enzymatic sensor for analytical detection [20].

Titanium oxide clusters—TOCs (also called polyoxotitanates or titanium–oxo clusters), belong to a class of molecular analogs of TiO_2 and titanate minerals, which have a well-defined structure and are passive in functionalization. TOCs have good chemical stability, photocatalytic properties, and redox activity, making them promising materials for developing hierarchical structures using bottom-up approaches [7]. Chen et al. [21] linked the TOCs-based tetrahedral clusters of Ti_4L_6 (L = embonate) with transition metal ions, such as Zn^{2+}, Co^{2+}, and Mn^{2+}. Ti_4L_6's tetrahedral geometry and its calixarene-like coordination active vertices facilitate the incorporation of coordination bond metal through π–π stacking interactions and other non-covalent interactions, which characterize it as an excellent building block. Metal–organic frameworks (MOFs) based on TOCs showed high water and heat resistance and excellent gas adsorption properties [21].

Zinc oxide is an n-type semiconductor, which has a wide band gap around 3.37 eV and high exciton binding energy (~60 meV). Due to its unique multifunctional physical and chemical properties, as well as its biological activity, it has called the attention of the scientific community a few years [22, 23]. Hassan and collaborators [24] developed a new nanomaterial from zinc oxide nanoparticles (ZnO NPs) microrecycled from spent Zn-C batteries, and 3D graphene hybrid aerogel. Graphene interacts weakly with ZnO NPs via van der Waals weak interaction, thus the electronic properties of graphene could be conserved in the hybrid material. The nanomaterial has elevated sensing properties with higher sensitivity, lower detection limit and fast response-recovery time, proving to be efficient for the detection of NO_2 due to the improvement in the conductivity of the nanomaterial and the porous structure of grapheme. It is interesting to point out that decorated ZnO NPs' gas detection mechanism follows the surface charge model [24].

2.3 Carbon-Based Materials

Carbon-based materials comprise an important class of materials used as molecular building blocks in supramolecular chemistry; this is mainly due to the wide variety of allotropes from 0 to 3D that provide numerous possibilities for constructing new supramolecular entities. They include graphene-based materials, fullerenes, carbon quantum dots, nano-onions, carbon nanotubes, nanocones called nanohorns, and nanodiamonds [25]. It is also important to emphasize that these materials mostly have low cost, good mechanical stability, flexibility, biocompatibility, in addition to being thermal and electrical conductors [26]. Figure 2 shows the structure of some carbon allotropes and their main properties.

Carbon dots are a newer class of carbon allotropes prepared by top-down and bottom-up approaches [8]. Due to excellent physico-chemical properties and biocompatibility, the carbon dots have aroused the scientific community's interest in recent years. It is noteworthy that with their ease of functionalization, they can be used as self-assembly building blocks [29]. Furthermore, carbon dots can be synthesized from abundant and non-toxic materials, such as biomass [30].

Sarkar et al. [29] developed carbon dots with different aliphatic/aromatic substitutions (cholesteryl, palmitoyl, naphthyl) with spacers, such as ethylenediamine, p-phenylenediamine, 2.2′-(ethylenedioxy) bis (ethylamine). Moreover, they demonstrated the formation of self-assemblies using amphiphiles exclusively based on carbon dots for the first time. Wang et al. [25] in turn, synthesized fluorescent carbon dots with [$GdCl_3Br$] with magneto-responsive properties (CQDGd) and used them

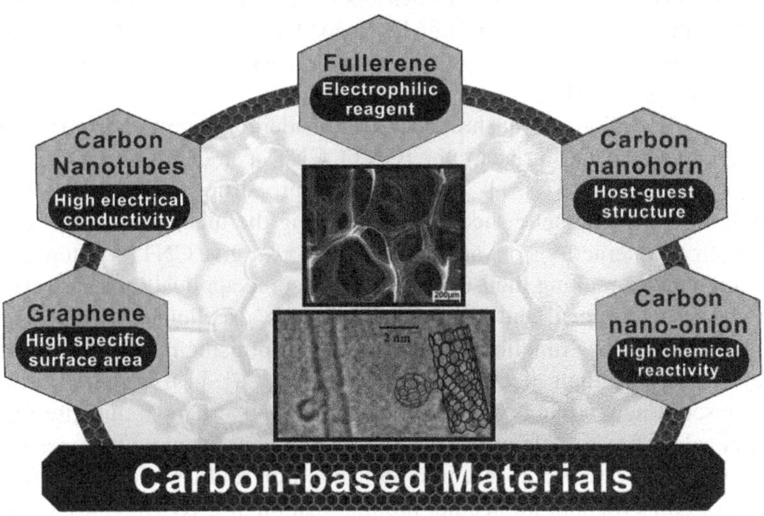

Fig. 2 Schematic representation of the main types Carbon-based materials. Adapted and reprinted with permission from Refs. [27, 28]. Copyright (2007 and 2012) American Chemical Society and Nature

as building blocks to form hybrid DNA nanospheres (ssDNA) using a simple self-assembly synthetic route. The phosphate and nitrogen bases of the dimple DNA strand served as self-assembly sites; thus, the CQDGd/ssDNA nanospheres were formed by interactions between Guanine and Cytosine through hydrogen bonds, hydrophobic interactions, and synergistic electrostatic forces [25].

Fullerenes are zero-dimensional (0D) carbon nanomaterials with hollow and perfectly spherical structures made up of five- and six-membered rings. There are several forms of fullerene, such as C_{60}, C_{70}, C_{76}, C_{82}, and C_{84}, with C_{60} being the most commonly studied [25, 31, 32]. These nanomaterials have a high affinity for electron transport [31] and have shown great potential in applications in areas, such as, energy and photocatalysis [32], solar cells, photodynamic therapy [33], biomedicine, among others. Shahzad et al. [34] used fullerene as a building block to develop a new heterogeneous photocatalyst for H_2 synthesis. Among the factors that contributed to the excellent performance of the photocatalyst, WO_3/fulereno@Ni_3B/$Ni(OH)_2$, the synergistic effect between the WO_3/fullerene nanocomposite and the cocatalyst, Ni_3B/$Ni(OH)_2$ stood out [34].

The combination of mechanical, thermal, optical, and electrical properties make graphene, a 2D (two-dimensional) material, the most widely investigated carbon-based material [26, 35]. Graphene has a unique crystal structure, having carbon atoms bonded in sp^2 hybridization, organized as a flat sheet of one atom thick [35]. One of their derived is graphene oxide which there is the presence of oxygenated groups, such as hydroxyls or carbonyls, in its structure. In recent years, graphene oxide, GO, has been used as a two-dimensional building block to form supramolecular hydrogels. Liu et al. [36] used graphene oxide to form hybrid supramolecular hydrogels with amino-functionalized silica nanoparticles (SiO_2–NH_2 NPs). Herein, SiO_2–NH_2 NPs interacted with GO nanosheets through hydrogen bonds between –NH_2 and –COOH in SiO_2–NH_2 and GO, respectively forming insoluble nano building blocks in a wide concentration range [36].

Another carbon allotrope that has received considerable attention from researchers in recent years is the carbon nanotube (CNT). CNTs can exhibit one, two, or more tubular graphite walls [31, 37]. The electronic properties and chirality of CNTs depend on how the graphene sheet is rolled during the tubular structure formation [37]. The strong interactions of van der Waals and π–π give CNTs a potent tendency to form aggregates, making their use difficult for some applications. However, their unique chemical, physical and electronic properties, excellent mechanical properties, and tunable surface functionality make CNTs promising in developing different materials, especially their electronic properties [38, 39]. In this sense, Ortiz et al. [39] proposed a new strategy for forming an innovative nanoarchitecture through the non-covalent functionalization of multi-walled nanotubes with ConcanavalinA (ConA), a protein, directly. ConA immobilization on multi-walled CNTs provided specific properties for supramolecular immobilization of glycoenzymes to construct bienzymatic glucose biosensors [39].

In the class of carbon allotropes, carbon dots have shown an upward trajectory for future applications due to luminescence properties, the possibility of green synthesis, and nanoscale dimensions. However, graphene-based materials represent the most studied carbon-based material, followed by carbon nanotubes until today [8].

2.4 Coordination Compounds

The directed synthesis of coordination compounds has attracted considerable attention in supramolecular chemistry and crystal engineering [4]. For instance, bis (benzimidazole) (Bbim) has been widely used in coordination chemistry, as it can join two or more metal centers, therefore, considered a ligand with variable denticity. Samoľová et al. [4] synthesized four new complexes with manganese (IV) and bis (benzimidazole) with different crystal structures in all complexes demonstrating that the Bbim ligand is a robust building block in the formation of supramolecular complexes [4]. Sun et al. [5] used polydiallyl malonylurea (PDAMU) as a building block to form complexes with silver ions (PDAMU-Ag). The PDAMU-Ag complex exhibited high antibacterial activity against Gram-positive bacteria (*S. aureus*) and Gram-negative bacteria (*E. coli*). Other results suggest that PDAMU can be used as a tool for constructing organometallic systems contributing to supramolecular chemistry and materials science [5].

The Prussian Blue (PB) and Prussian Blue Analog (PBA) complexes are important building blocks for the synthesis of different nanocomposites. These compounds have structural, thermal, and electrochemical properties which make them promising candidates for applications in different areas. PB and PBA present the general structural formula $M_{ax}{}^{+}[M_b(CN)_6]_y \cdot nH_2O$, in which, for PB, $M_a = M_b =$ Fe, and for PBA, M_a and M_b represent a transition element [40, 41]. It should be noted that PBA-based materials are low cost and easy to obtain [42]. Niu et al. [42] developed a Prussian blue analogue of nickel and cobalt and used it as a new high-efficiency non-enzymatic sensor for glucose detection. The developed Ni-Co PBA presented an excellent catalytic performance for the electro-oxidation of glucose when compared with pure Ni PBA and pure Co PBA. The analysis of the experimental data confirmed that the sensor based on Ni-Co PBA had a low detection limit (1.2 m mol L^{-1}), wide linear range (0.002–3.79 m mol L^{-1}), good selectivity and high sensitivity (149 μA mM^{-1} cm^{-2}) suggesting that this is a promising sensor for detecting glucose in real samples [42].

Rodrigues and collaborators [41] used the Prussian blue analogue of copper ($Cu_2[Fe(CN)_6]$) as components for the formation of the $CuO/Cu_2[Fe(CN)_6]$ nanocomposite and investigated its electrical properties in the presence of H_2O_2. The data obtained by cyclic voltammetry revealed a change in faradaic currents in the presence of peroxide; the nanocomposite showed two well-defined redox pairs (from Cu^{+}/Cu^{2+} and $CuFe^{2+}$ to $CuFe^{3+}$), on the other hand, the precursors, CuO and $Cu_2[Fe(CN)_6]$, showed only a slight intensified effect for oxidation [41].

Carvalho et al. [43] used the cobalt Prussian blue analogue, $Co_3[Co(CN)_6]_2$, to construct a magnetically separable nanocomposite through interaction with magnetic nanoparticles. The developed $Co_3[Co(CN)_6]_2/Fe_3O_4$ nanocomposite presented a nanocubic shape decorated with small magnetite spheres that interacted through dipole–dipole interactions. An increase in faradaic currents for the Co^{2+}/Co^{3+} pair present in the voltammograms of the hybrid nanomaterial compared to the isolated, $Co_3[Co(CN)_6]_2$ was observed. The magnetic properties of Fe_3O_4 NPs were preserved even after adsorption of cobalt Prussian blue analogue nanocubes. The nanocomposite also showed promising catalytic properties, demonstrating, therefore, that the approach used is effective to produce hybrid nanomaterials with a combination of magnetic and electrochemical properties [43].

Moraes and collaborators [44] developed a new hybrid material consisting of Prussian blue (PB) and cashew gum polysaccharide (CG), a biopolymer that has antibacterial, antitumor, healing, and gastric potential properties, in addition to being biocompatible, biodegradable, and non-toxic and determined its potential as an electrochemical sensor for the oxidation of drugs. The analysis of the data obtained by the electrochemical tests showed that the bionanocomposite formed, PBNPs@GC, has the capacity to oxidize some drugs, such as acetaminophen (ACT), metamizole (MTM), and methotrexate (MTX), indicating that this material shows promise for the development of sensors for clinical and pharmaceutical samples [44].

2.5 Others Classes of (Nano)materials

Another material class which is commonly used as a building block for supramolecular chemistry is polyphenols (tannins), which has several advantages for engineering biomaterials, as they are biodegradable, biocompatible, and abundant in nature. The properties of these biomaterials provide the possibility of use as antioxidant, drug delivery system, cell biohybrids, and for diagnosis [45]. The natural polyphenol, tannic acid (TA), and poly(2-n-propyl-2-oxazoline) (PnPropOx) were used by Mathivanan et al. [46] for the formation of TA/PnPropOx multilayer thin films joined by hydrogen bonding and hollow capsules manufactured by a sequential coating of TA and PnPropOx onto $CaCO_3$. The authors further investigated the interactive forces, stability, permeability, and film thickness. It was found that the multilayer thin films and capsules obtained were considered stable over a wide pH range (2–9) and that hydrogen bonds and hydrophobic interactions are responsible for improving the stability of the capsules [46].

Some biocompatible molecules, such as peptides, copolymers, phospholipids, and small synthetic molecules have evolved as building blocks with great potential for assembly into nanostructures with controllable size, shape, charge, and surface properties. These molecules can spontaneously assemble, forming ordered superstructures promising for applications in some areas, such as, energy, catalysis, optics, medicine, among others [9]. Long et al. [9] showed the recent advances in the development of trigonal building blocks in their work. Due to the unique structure of C_3 symmetry, molecules with trigonal geometry present advantages for their use in supramolecular self-assembly as their structure expands the variety of building blocks of supramolecular materials. Trigonal building blocks are generally small and easily obtainable molecules; in addition, they are more sensitive to stimuli than polymers, for example, which are structurally complex [9]. Trigonal molecules can be synthetic molecules [47, 48], nucleic acids [49], and peptides [50]. It should be noted that the peptides, being biodegradable and biocompatible, are attractive as building blocks for supramolecular biomaterials and can be used in different areas, such as biotechnology and biomedicine [50].

In addition to the carbon dots presented earlier, other quantum dot (QDs) materials have attracted the interest of the scientific community over the years mainly due to their tunable bandgaps and the possibility of functionalization, thus, making them efficient building blocks for many optoelectronic applications [51, 52]. Sol et al. [53] developed a Z-type composite photocatalyst based on TiO_2 nanorods decorated with of gC_3N_4 and Au quantum dots. The Z-type heterojunction formed at the r-TiO_2 and gC_3N_4 interface as well as the improved electron capture by the Au QDs were responsible for the improvement in the catalytic performance of the composite photocatalyst. Therefore, this work provides an efficient strategy to design high-performance photocatalysts using quantum dots and nanorods as building blocks [53].

Among the fields of supramolecular chemistry, "host–guest chemistry" performs a significant role in forming new supramolecular entities and has come to the scientific community's attention over the years. In host–guest chemistry, small molecules (guest) are reversibly encapsulated in the cavities of larger molecules (host) [54]. Cyclodextrins (CDs) [55], cucurbit[n]urils (CB[n]s) [56], calix[n]arenes (CA[n]) [54], and pillar[n]arenes (PA[n]) [54] are among the most used molecules as "host molecules". In 2019, Yao et al. [57] obtained a host–guest complex with β-cyclodextrin and methylene blue in a spray-coating based on hyaluronic acid. The obtained complex showed a good photodynamic antibacterial capacity and can be used to sterilize medical devices [57]. The tin(IV) porphyrin, viologen, and cucurbit[8]uril (CB[8]) complex developed by Shee et al. [56] can be utilized as a supramolecular sensor for the detection of aromatic compounds (hydroquinone, naphthalene). IN this work, the authors used a tin(IV)porphyrin functionalized with viologen "as a guest molecule" since the viologen units can be encapsulated within the CB cavity. In this case, the porphyrin served as a structural scaffold for the complexation host–guest [56].

Tang et al. [58] developed ferrocene-functionalized poly(ionic liquid) microgel nanoparticles (PIL-Fc NPs). The possibility of using PIL-Fc NPs as a building block for self-assembly was evaluated and confirmed by forming an inclusion complex between the obtained nanoparticles and the β-cyclodextrin dimer (β-CD). The supramolecular self-assembly was performed through the host–guest interaction of the ferrocene groups, present in the functionalized poly(ionic liquid) microgels, with the β-CD dimer [58].

3 Methods of Building Blocks Supramolecular Assembly onto Electrode Surfaces

The successful applications in the supramolecular electrochemistry are dependent on the deposition of the building blocks onto electrode surfaces. The supramolecular immobilization processes can improve molecular recognition, providing superior electrochemical stability and enhance electron transfer [59]. The supramolecular electrochemistry mechanism is guided by interactions that occur in the electrode interface [60]. Therefore, the supramolecular electrochemistry investigate the interactions at the molecular level between building blocks and electrode surface [60]. Furthermore, the reactivity of the interface constituted by site-specific supramolecular interaction/analyte are objects of study of supramolecular electrochemistry. Thus, the formed supramolecular structure onto electrode surface is influenced by the balance of all interactions that occur between the building blocks and the analyte (metals, ions, molecules, metabolites, proteins, hormones, nucleic acids, etc.) [60]. For example, hydrogen-bonding, electrostatic interactions and halogen-bonding (supramolecular anion receptors) can promote the anion recognition to improve their electrochemical sensing [61]. Moreover, supramolecular self-assembly of building blocks onto electrode surfaces exhibit high molecular recognition capability (via host–guest interaction, hydrogen-bonding and π–π interactions) that can be improve electrochemical performance of (bio)sensors [39, 62].

In the development these supramolecular structures, building blocks are immobilized onto electrode surfaces using various strategies [63–66]. Figure 3 presents a schematic representation of the main strategies used for supramolecular assembly of building blocks onto electrode surfaces.

These assembly methods are important because they have the capacity of form self-organized supramolecular architectures onto electrode surface. Table 1 presents the main advantages, disadvantages and suggested interaction in the electrochemical interface to form self-organized supramolecular structures onto electrodes surfaces. The self-assembled monolayes (SAMs), LbL self-assembly and electrochemical deposition methods are described with more detail in the following subsections.

3.1 Self-Assembled Monolayes (SAMs)

The SAMs method is one of the most used strategies for preparation of organic ultrathin films with controlled thickness and chemical functionalization. The self-assembly process is governed by chemical interaction of alkanethiols, dialkyl disulfides, alkylalkoxysilanes or sulfides with the surface of metallic or semiconductor electrodes (gold, copper, mercury, platinum, silver or silicon, indium tin oxide) [64, 67]. Various types of interactions can occur in the electrode interface include

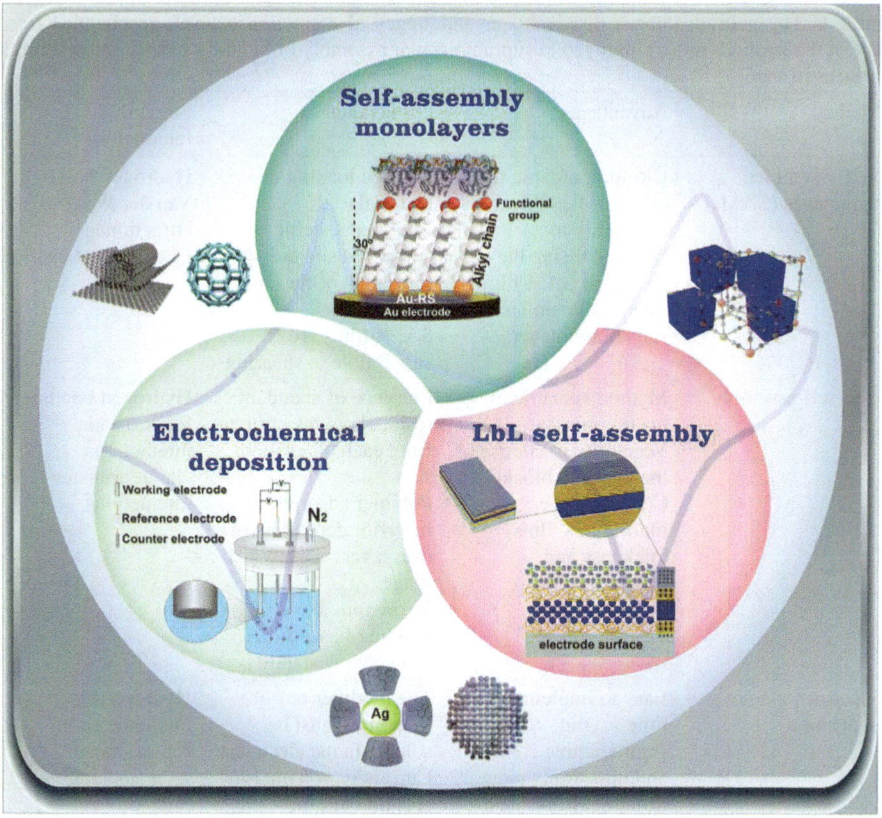

Fig. 3 Schematic representation of the main methods used to supramolecular assembly of building blocks onto electrode surfaces

chain-chain and non-bonded interactions (interchain van der Waals, steric, repulsive, and electrostatic forces), which ensure tight packing, stability and organization of the organic monolayer [64]. Moreover, SAMs nanoarchitectures form positive/negative/hydrophobic terminal chemical groups (–COOH, –OH, $–NH_2$, –CN, –SH, $–SO_3H$, –SiOH, CH_3) that can interact with others chemical species in the interface of the modified electrode [64, 67]. SAMs approach involve the formation of a (supra)molecular structures that exhibited unique physical and chemical properties on electronic devices such as organic film solar cells, organic thin film transistors, electrochemical sensors and biosensors [68, 69].

The SAMs approach improve aspects related the electronic functions of the organic film-based devices to as the dominant mechanism of charge transport and electrical output [68–70]. For example, the insertion of SAMs structures in the electrode/semiconductor interface of organic transistors can control/improve the charge injection and transport [68, 70]. Furthermore, the SAMs (supra)molecular films play

Table 1 The main advantages, disadvantages and suggested interaction in the electrochemical interface of the methods more employed to supramolecular assembly of building blocks onto electrode surfaces [63–66]

Method	Advantages	Disadvantages	Suggested interactions
Self-assembled monolayes (SAMs)	Ultrathin organic films of controlled thickness Flexibility in functionalising the end groups of SAM films Thermodynamically stable monolayers	Limited loading of materials Required specific compounds to the formation of the monolayers SAMs films are susceptible to shearing	Hydrogen bonding Van der Waals attraction Chemical interaction
LbL self-assembly	Method versatile of operation easy Versatility to choose of the building blocks Control of the multilayers films thickness and physicochemical properties	Presence of abundant residual polyelectrolyte from each deposition step Polymer adsorption behavior dependent of over a very narrow pH range Hydrogen bonding-based films showed low stability	Hydrogen bonding Electrostatic interactions charge-transfer interactions
Electrochemical deposition	Easy to implement and time-saving Reproducible control and uniformity over film thickness Ability to use diverse electrode of complex geometries	The modifier or monomer must be soluble in the electrolyte Current or voltage is required to film synthesis Removal film onto electrode surface is difficult	Hydrophobic interaction Ion-exchange π–π stacking interaction

an important role in the tuning of electronic, structural and morphological properties of the interface of molecular devices. Lin et al. (2020) examined the electric behavior of bulk-heterojunction (BHJ) organic photovoltaic (OPV) cells based in SAMs composed of [2-(9H-Carbazol-9-yl)ethyl]phosphonic acid (2PACz) [71]. Self-assembly molecules of 2PACz onto the indium tin oxide (ITO) surface were used as a hole-selective interlayer to enhance the power conversion efficiency and operational stability of ITO-2PACz. This behavior was because of the decrease contact-resistance, enhanced charge extraction in the ITO-2PACz/BHJ interface, high charge-carrier mobility before and after photoaging [71]. Very recently, it was reported a novel approach that involves interstitially mixed self-assembled monolayers (imSAMs) from a molecular diode of 2,2′-bipyridyl-terminated *n*-undecanethiol ($HSC_{11}BIPY$) and a nonrectifying diluent of *n*-octanethiol (HSC_8) onto gold substrates [72]. These methods consist in repeated surface exchange of molecules (ReSEM), resulting in filling of HSC_8 into interstices formed by bulky $HSC_{11}BIPY$, decreasing the defects within the monolayers [72]. Thus, the supramolecular packing structure of the SAMs enhanced electrical stability, functionalities and yields of working junctions.

Another important characteristic widely exploited of SAMs supramolecular architectures is the control and guidance of the terminal chemical groups that can specifically interact with analyte of interest [73–75]. The molecular interface based in the host–guest interaction was manufactured from cucurbit[7]uril (CB[7]) on mixed ferrocenylundecanethiolate/n-alkanethiolate SAMs onto gold electrode [73]. As a result, the supramolecular hosts assemblies at the nanometer scale demonstrated the capability of forming inclusion complexes (CB[7]@Fc) to probe the structural heterogeneity of ferrocenyl (Fc) SAMs [73]. Specific interactions were used by Li et al. (2019) to immobilize carboxyl-functionalized graphene (CFG) onto glassy carbon electrode (GCE) surface [74]. In this study, SAMs method was developed to promote the covalent interaction between ethylenediamine ($NHCH_2CH_2NH$) and CFG. The functionalized electrode (CFG-$NHCH_2CH_2NH$/CGE) SAMs-based was crucial to electrochemical detection of hydroxyl free radicals [74].

3.2 *LbL Self-Assembled*

Among the self-assembled building blocks-based methods, LbL self-assembled is one of most used strategies to make multilayer thin films with controlled thickness on a variety of electrode surfaces [76, 77]. The principle behind the of formation of multilayered films are based in various types of non-covalent and covalent interactions (e.g., electrostatic interactions, hydrogen bonding, charge-transfer interactions, coordination interactions, sol–gel processes, hydrophobic interactions, etc.) [78]. Consequently, LbL method is a highly versatile technique that exhibit different experimental approaches, such as dipping LbL assembly, spin-assisted LbL assembly and spray-assisted LbL assembly [79]. Among them, dipping LbL coating is the most used method to the produce functional supramolecular structures. The dipping LbL approach consists in alternating immersion of a charged conductive substrate into aqueous solutions of oppositely charged building blocks [76–79]. Then, rinsing the as-prepared films in appropriate solutions and drying under nitrogen atmosphere for the removal of weakly adsorbed species occurs as intermediate steps. In this assembly process, thickness and physicochemical properties of thin films can be controlled with precision through of adjustment of the experimental conditions including solution pH, ionic strength, addition of salt and polyelectrolyte charge density etc. [65, 76]. Based on this information, LbL method is one of the most easy and versatile strategies to supramolecular self-assembly of functional building blocks onto electrode surfaces. Therefore, the LbL self-assembly allows investigate and control supramolecular specific interactions that occurs in electrochemical interface constituted by building blocks and chemical species of interest. The self-assembly of building blocks in electrodes with desirable architectures and properties is of great importance for study the performance of electronic and electrochemical devices.

Regarding electrochemical sensors and biosensors, LbL method promote the improve in the electronic conductivity, and electroactive surface area, enabling an increase in sensitivity [80–83]. However, the most important aspect is the introduction and control of sensitive molecular recognition species in electrode surface that improve the selectivity capability in complex samples [81, 82]. For this purpose chemical groups can be anchored to interact specifically with target analytes through non-covalent and covalent interactions (e.g., host–guest interaction, electrostatic interactions, biospecific interactions, chiral recognition, "click chemistry" reaction, etc.) [78, 83]. A supramolecular electrochemical architecture with molecular recognition properties was prepared by LbL self-assembly of polyethylene imine-reduced graphene oxide (PEI-rGO) composites and cholesterol oxidase (ChOx) [81]. In this work, the electrostatic interaction between the positively charged PEI-rGO and negatively charged ChOx increase the electron transfer ability between the redox enzyme and the surface of glassy carbon electrode (GCE). This supramolecular interaction that occur at the electrochemical interface improve the electric conductivity and maintain the bioactivity of the ChOx [81]. Another example of supramolecular multilayers LbL strategy-based with biorecognition component (folic acid, FA) was demonstrated by Correia and colleagues [82]. The authors realized alternately deposition of positively charged polyallylamine hydrochloride (PAH) and negatively charged FA on indium tin oxide (ITO) electrodes. The multilayers structure increase electroactive surface area and improve sensitivity/selectivity to electrochemical detection of folate receptor alpha (FR-α) via binding affinity between FA and FR-α [82].

The LbL method allows study detailed electrochemical mechanisms and electrochemical reactions that occur on interface of electrodes used in energy conversion and storage devices [84]. Thus, the possibility of control the thickness and architecture of multilayed electrodes in nanoscale is crucial to understand the processes that involves competition between mass and charge transfer, modulation of interfacial dipole, and ion permeability [84]. Wang et al. (2021) developed LbL self-assembled nanostructured electrodes to investigate the electrochemical performance of lithium-ion batteries (LIBs) [85]. The electrostatic interaction and colloidal stability properties of building blocks (nanoparticles of lithium titanate, lithium iron phosphate, lithium manganese oxide, and carboxymethyl celulose) were useful for make LIBs based in the LbL self-assembly. The supramolecular interactions improved Li^+ intercalating phase and electron-conducting phase providing a high specific capacity during the charge/discharge cycles [85]. In another work, researchers explored the spray-assisted LbL assembly technique to make nanostructured electrodes and study their electrochemical behavior as flexible solid-state supercapacitor [86]. In this case, the method is based in alternating spray of graphene quantum dot-reduced graphene oxide (GQD-rGO) solution and an aqueous polyaniline (PANI) solution onto flexible conductive substrate of carbon fiber cloth (CFC) [86]. The spraying-based self-assembly process of GQD-rGO changed the CFC surface from hydrophobic to hydrophilic and enhanced the electrostatic interaction between PANI and CFC. The supramolecular strong interactions between GQD-rGO and PANI in the CFC electrode interface improved the cycling stability and specific capacitance [86].

3.3 Electrochemical Deposition

Compared with the approaches presente in Table 1, electrochemical deposition methods are most inexpensive with easy operational implement and time-saving, also showing reproducible and uniform film structrures onto various electrode geometries [59, 63, 66]. Because of these characteristics, the electrochemical deposition is extensively used to supramolecular self-assembly of building blocks on the conducting electrode surfaces. The electrochemical deposition has as principle the immersion of three electrodes (working, reference, and auxiliary) in suitable electrolyte solution where sufficient current or voltage is provided to reduce ionic metallic species and oxidize or reduce a monomer to polymerize it on the surface of the working electrode [59, 63, 66]. For this reason, the main approaches of electrochemical deposition are the electrodeposition of metal and metal-based (nano)materials and electropolymerization of polymeric species. The experimental and electrochemical parameters such as current densities/voltages, substrate surface, electrolyte pH, solution viscosity, and diffusion coefficient can be adjusted and play critical roles in the morphology thickness, uniformity and reactivity of the deposited film [59, 87]. Therefore, electrodeposition and electropolymerization are approaches of supramolecular electrochemistry to investigate the effects of electrodeposited films in the electrochemical behavior of (bio)sensors, corrosion protection, energy conversion and storage [88–96].

In the context of (bio)sensors, electrochemical deposition methods enhanced sensitivity and selectivity through of the supramolecular interactions of (bio)molecules with the electrochemical interface [88–90]. In this process, (supra)molecular recognition and (bio)recognition ability takes effect in the electrochemical behavior of (bio)sensors because of the host–guest interactions (inclusion complexes), hydrogen bonds, electrostatic interactions and halogen-bonding [88–92]. For example, cyclodextrins-based host–guest complexes confine the target analyte next of the electrode surface, providing a decrease or increase of the electric current of analytes' redox processes [89]. The electrodeposition was used as strategy of supramolecular electrochemical sensing by Uppachai and coworkers [91]. One of the first steps was electrodeposition of Au nanoparticles onto graphene oxide modified glassy carbon electrode (AuNPs/GO/GCE). To improve the detection of dopamine (DA), self-assembled mixed surfactants (TBABr/SDS) were added into electrolyte solution [91]. Thus, hydrophobic interaction and electrostatic attraction occur through C–H chain and charge terminal group of the TBABr/SDS with DA structure, respectively. These supramolecular interactions and synergistic effect of TBABr/SDS with AuNPs/GO provide an enhanced anodic peak current of DA [91]. Very recently, it was reported a novel approach based on preconcentration of the organic target analyte using the electropolymerization [92]. The poly-(para-aminophenol) film was electrochemically deposited in the glassy carbon electrode (GCE) surface. The specific interactions that occur in the CGE/OxPPAP electrode interface are important to elucidate the electrochemical mechanism in the detection the melamine (Mel). As a result, the covalent supramolecular interaction (Mel-aldehyde resins) between melamine and the aldehyde-rich CGE/OxPPAP surface increase sensitivity and selectivity to determine Mel in real milk samples [92].

Concerning others applications (anticorrosion, energy conversion and storage), electrochemical deposition approaches can influence in the supramolecular organization, electrochemical properties and performance of the building blocks on the electrode surfaces [93–96]. In this context, it was demonstrated the electrodeposition of graphene oxide (GO) from covalent interaction with electroactive monomer (carbazole, Cbz) [93]. The electrochemical step is governed by electropolymerization of Cbz that covalently bond to GO surface to electrodeposit the GO-Cbz structures onto indium tin oxide (ITO) surface. The supramolecular self-assembly of GO-Cbz films in the conductive regions of the electrode enabled excellent anticorrosion performance [93]. The electropolymerization of poly-1,4-bis(2-thienyl)benzene (PDTB) film was used to prepared a photocathode and investigate the processes of photoelectrochemical (PEC) hydrogen evolution [95]. Electrosynthesis resulted in the PDTB-base photocathodes with various microporous structures and regulated S oxidation states. This structural organization improve the light absorption capacity, and charge separation of the photocathodes. Thus, self-assembly of microporous structures by electropolymerization promoted narrow band gap and low charge transfer resistance that facilitated the transport of photogenerated electrons, resulting in the enhance hydrogen production potential [95].

4 Applications of Supramolecular Structures

Nanotechnology has attracted tremendous attention in the past years from several areas of knowledge, mainly due to the unique properties and prominent advantages of nanoscale materials and devices [97]. More recently, there has been an extensive use of nanotechnology to develop new functional systems used as sensors, biosensors, molecular recognition devices, energy storage, and much more, as it is schematic represented in Fig. 4 [98]. Electrochemistry plays an important hole on the application of nanoscale materials in different areas, ranging from catalysis and energy to (bio)sensors development, and beyond [99]. The synergism between nanomaterial properties (already mentioned earlier in this chapter) and electrochemical devices, makes them a powerful tool for several applications, especially owing to the high sensitivity, portability, low cost, simplicity of instrumentation, and versatility of such devices [100].

Electrochemical systems have been coupled with different types of nanomaterials like graphene and its derivatives, nanoparticles, etc., with an increasing range of possibilities [100]. In this context, the use of self-assembled building blocks as modifiers onto the surface of electrodes can be an interesting approach to improve the conductivity and other electrical and chemical properties of electrodes for a given

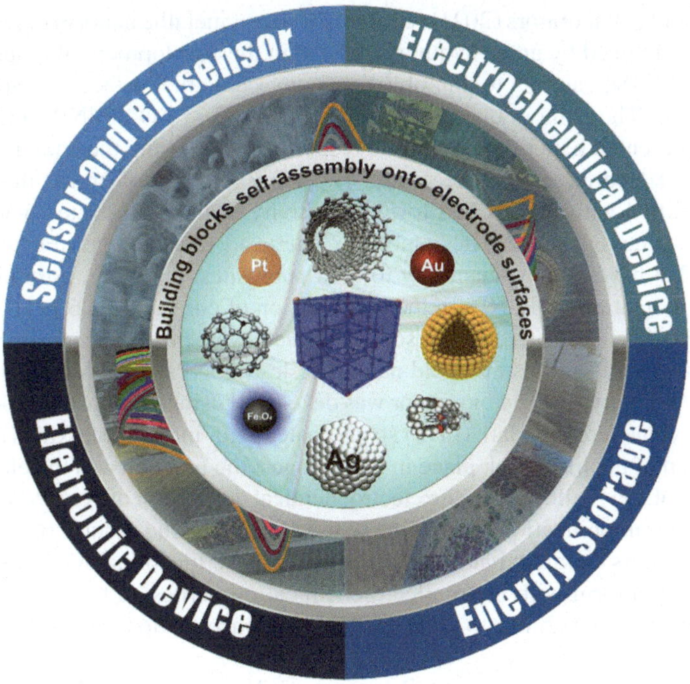

Fig. 4 Schematic illustration of the main applications of self-assembled building blocks onto electrode surfaces

application. Some interesting works involving supramolecular electrochemistry will be described as follows.

Roushani et al. (2020) described a nanoaptasensor to detect traces of an antibiotic commonly used in the treatment of bacterial infections called chloramphenicol (CAP). The device was constructed using graphene oxide functionalized with silver nanoparticles and (3-aminopropyl) triethoxysilane bounded to an aptamer (Apt/AgNPs/[NH_2-Si]-f-GO) modifying a glassy carbon electrode (GCE). 5×10^{-3} mol L^{-1} of $[Fe(CN)_6]^{3-/4-}$ solution was employed as a redox probe in this approach. The high sensitivity and good linear concentration range of the nanoaptasensor is attributed to the presence of the AgNPs/[NH_2-Si]-f-GO layer. This layer has a large surface area on which more Apt can interact through covalent bonding, resulting in more active sites for the biorecognition event. The authors demonstrated an indirect mechanism for the detection of CAP in which the interaction between the electrode modifier and the analyte (that is, the biorecognition event) creates a physical barrier/restriction, hindering the redox reaction of the probe on the electrode surface, decreasing the peak current. The proposed device has several advantages, including low detection limit, sensitivity, adequate stability, and high selectivity, in addition to conjugation with biomolecules [101].

Gayda and collaborators (2021) reported the use of metallic nanoparticles of Palladium (Pd), obtained by green synthesis (gNPs) in the development of amperometric biosensors (ABSs), aiming to immobilize enzymes onto the surface of graphite electrodes (GEs). The authors claim that two bioelectrodes, AO-gPdNPs and laccase-gPdNPs, presented better amperometric responses when using the enzymes alcohol oxidase (AO) and laccase, respectively. The study also demonstrated that bioelectrodes modified with PdNPs had lower sensitivity to their substrates, wider linear ranges of detection and greater stability. These are desirable characteristics for enzymes that have a high sensitivity to their substrates, therefore, pointing out the promising potential of gPdNPs in the construction of enzymatic ABSs [102].

Recently, Avelino et al. (2021) described the achievement of a biosensor composed of polypyrrole (PPy) films and gold nanoparticles (AuNPs) used for specific detection of HPV genotypes. The biosensor was developed using flexible electrodes based on polyethylene terephthalate (PET) strips coated with indium tin oxide (ITO). The construction of the films took place in three stages: the first one was the electropolymerization of PPy, followed by the electrochemical deposition of AuNPs and the last one was the immobilization of cysteine (Cys) on the metallic surface by chemisorption. The authors claimed that the association of these functional materials allowed obtaining a platform with better electrochemical properties and high surface area, besides the excellent ability to identify minimal concentrations of specific types of HPV [103].

The same group reported a similar approach but focused on development of a genetic device for the assessment and monitoring of human infection by SARS-CoV-2 in the early, intermediate, and late stages of the disease. The nanostructured platform was built with polypyrrole (PPy) and gold nanoparticles (GNP) on miniaturized gold-doped indium tin oxide (ITO) electrodes, functionalized with oligonucleotide. The electrode was built through electrochemical polymerization, followed by an electrochemical deposition of GNPs, and finally, a chemisorption of Cys molecules in GNPs was performed. The biosensor exhibited high selectivity by not recognizing the biological target in samples from patients not infected with SARS-CoV-2. The proposed device obtained an estimated linear response range from 800 to 4000 μL^{-1} copies with a regression coefficient of 0.99 and a detection limit of 258.01 μL^{-1} copies, proving to be efficient in the diagnosis of SARS-CoV-2 [104].

Metallic semiconductor oxides, used as building blocks in electrodes, have gained prominence in the literature, due to the possibility of modifying amperometric, potentiometric and conductometric properties. Fatema and Oh (2021) reported the obtaining of graphene-based electrodes modified with mesoporous oxides of Zinc (Zr), Silicon (Si) and Indium (In) (ZrO_2–Ag–G–SiO_2 and In_2O_3–G–SiO_2), for detection of non-enzymatic glucose, ascorbic acid, and albumin in urine at physiological pH. The authors stated that the In_2O_3–G–SiO_2 electrode performed well over a wide range of urine electrolytes, and showed no activity against uric acid, suggesting a potential for biodetection of glucose in urine [105].

Hashemi et al. (2020) proposed the modification of graphene oxide (GO) with Ag and Iron Oxide (Ag–Fe_3O_4) hybrid metallic nanoparticles for real-time detection of ascorbic acid (AA) in plasma blood samples through electrochemical measurements.

The results showed that the modification of glassy carbon electrode (GCE) with such hybrid material significantly improved its sensitivity and selectivity. The authors further stated that the GCE modified with GO–Ag–Fe_3O_4 had detection limits and sensitivity of 74 × 10^{-9} mol L^{-1} and 1146.8 μA mM^{-1} cm^{-2}, respectively, within the concentration range of 0.2–60 μM. In addition, the modified electrode maintained 91.23% of its total performance after 15 days of detecting the AA oxidation peak, highlighting its stability/selectivity, and evidencing the notable application capability of the proposed platform [106].

Carbon-based materials such as carbon nanotubes (CNTs), graphene oxide (GO) with different shapes and sizes have been widely used for applications such as sensors, biosensors mainly due to their excellent electrical properties, as already stated earlier in this chapter [98]. Xia et al. (2022) obtained a biosensor (GOx/β-CD/MWCNTs/GCE) constructed from a graphite carbon electrode (GCE) modified with multiwalled carbon nanotubes (MWCNTs) functionalized with β-cyclodextrin and glucose oxidase (GOx), for the detection of glucose in honey. The study stated that the interaction of β-cyclodextrin (β-CD) with CNTs improved the dispersibility of the nanotubes in aqueous media, due to the internal hydrophobicity and external hydrophilicity of β-CD. The results obtained in this study revealed that the carbon nanotubes promoted the direct transfer of electrons from GOx, showing a linear response for glucose concentration in the range of 50 μM to 1.15 mM with a detection limit of 0.42 μM and a sensitivity of 32.28 μA mM^{-1} cm^{-2} [107].

It is important to highlight that the manipulation of matter at the nanoscale also enabled the construction of electrical devices used as capacitors and batteries for energy storage. Zamiri et al. (2021) synthesized a ternary nanocomposite consisting of three-dimensional graphene, silver nanoparticles and polyaniline (3DG-nAg-PANi) using an in-situ polymerization technique, for applications as asymmetric supercapacitor (ASC). The study showed that the incorporation of PANi with 3DG led to the rapid diffusion of Ag ions, significantly improving the conductivity of the nanocomposite, and increasing the peak area of the cyclic voltammogram. Therefore, the authors concluded that the synergistic combination of nanomaterials (3DG-nAg-PANi) improved the charge storage property of the electrodes, showing an increase of almost 2.5 times in capacitance when compared to 3DG/PANi composite, highlighting a potential use of such material for future energy storage devices [108].

Coordination compounds, such as Prussian Blue, has also been applied in development of electrochemical nanosystems (e.g., energy storage devices and (bio)sensors) for quite some time, with remarkable success reported in the literature over the last years [109]. As an example, Carvalho and Collaborators (2021) described recently the development of a carbon nanotube-based material (SWCNT-COOH) coupled with Prussian Blue nanoparticles (PBNPs) as an promising electrode modifier for the indirect detection of Ibuprofen (IBP) [110]. The electrode was built through the incorporation of PBNPs in the internal and external surface of the nanotube. Cyclic voltammetry (CV) data revealed that the presence of IBP decreases the electrochemical currents of the redox processes of PBNPs, featuring an inhibition effect that is proportional to the analyte concentration, suggesting an interaction between IBP and free Fe^{3+} sites in the complex structure. Thus, IBP could be detected indirectly at lower potential when compared to the direct IBP oxidation potential. This electrochemical behavior opens possibilities for future studies of IBP detection based on the inhibition effect of this analyte on the redox processes of PBNPs.

5 Conclusions

This chapter describes a brief electrochemistry supramolecular approach of smart self-assembled nanostructures constructed by appropriate choice of building blocks, synthetic route, and immobilization methods. The combined effect of lower-level components, that could only be observed at the nanoscale, point to another avenue to pursue within the constitutional dynamic chemistry. Special attention is given to choosing the electrochemical technique in order to provide information on energy and kinetics that is not available with spectroscopic and mass spectrometric techniques. Based on the studies summarized in this chapter, the nano-supramolecular assembled nanomaterials present the potential for various studies and applications including drug carriers, (bio)sensor, tumor diagnosis, electronic device, and energy storage.

Acknowledgements The financial support from CAPES and CNPq (436086/2018-2 and 314456/2020-1 projects) is gratefully acknowledged.

References

1. Silva, A.T.B., et al.: Nano-assembled supramolecular films from chitosan-stabilized gold nanoparticles and cobalt(II) phthalocyanine. J. Braz. Chem. Soc. **24**, 1237–1245 (2013)
2. Carvalho, C.L.C., et al.: New hybrid nanomaterial based on self-assembly of cyclodextrins and cobalt Prussian blue analogue nanocubes. Int. J. Mol. Sci. **16**, 14594–14607 (2015)
3. Zhang, T., Ran, J., Ma, C., Yang, B.: A universal approach to enhance glucose biosensor performance by building blocks of Au nanoparticles. Adv. Mater. Interfaces **7**, 1–9 (2020)
4. Samoľová, E., Premužić, D., Plociennik, S., Hołyńska, M.: Bis(benzimidazole) as supramolecular building block in manganese(IV) chemistry. J. Mol. Struct. **1176**, 366–375 (2019)
5. Sun, M. et al.: Multifunctional polymer bearing malonylurea groups for the fabrication of coordination complexes and supramolecular assemblies. Eur. Polym. J. **156**, 110616 (2021)
6. Pinyou, P., Blay, V., Muresan, L.M., Noguer, T.: Enzyme-modified electrodes for biosensors and biofuel cells. Mater. Horizons **6**, 1336–1358 (2019)
7. Liu, C., Wang, Y.: Supramolecular chemistry of titanium oxide clusters. Chem. A Eur. J. **27**, 4270–4282 (2021)
8. Zhou, Y., Mintz, K.J., Sharma, S.K., Leblanc, R.M.: Carbon dots: diverse preparation, application, and perspective in surface chemistry. Langmuir **35**, 9115–9132 (2019)
9. Long, K., Liu, Y., Li, Y., Wang, W.: Self-assembly of trigonal building blocks into nanostructures: molecular design and biomedical applications. J. Mater. Chem. B **8**, 6739–6752 (2020)
10. Lindoy, L.F., Park, K.M., Lee, S.S.: Metals, macrocycles and molecular assemblies – macrocyclic complexes in metallo-supramolecular chemistry. Chem. Soc. Rev. **42**, 1713–1727 (2013)

11. Manke, A.M., Geisel, K., Fetzer, A., Kurz, P.: A water-soluble tin(iv) porphyrin as a bioinspired photosensitiser for light-driven proton-reduction. Phys. Chem. Chem. Phys. **16**, 12029–12042 (2014)
12. Gouyau, J., Duval, R.E., Boudier, A., Lamouroux, E.: Investigation of nanoparticle metallic core antibacterial activity: gold and silver nanoparticles against escherichia coli and staphylococcus aureus. Int. J. Mol. Sci. **22**, 1–15 (2021)
13. Mosquera, J., et al.: Plasmonic nanoparticles with supramolecular recognition. Adv. Funct. Mater. **30**, 1–17 (2020)
14. Montes-García, V. et al.: Harnessing selectivity and sensitivity in ion sensing via supramolecular recognition: a 3D hybrid gold nanoparticle network chemiresistor. Adv. Funct. Mater. **31** (2021)
15. Golub, E., Pelossof, G., Freeman, R., Zhang, H., Willner, I.: Electrochemical, photoelectrochemical, and surface plasmon resonance detection of cocaine using supramolecular aptamer complexes and metallic or semiconductor nanoparticles. Anal. Chem. **81**, 9291–9298 (2009)
16. Lopes, L.C.S., et al.: Silver and gold nanoparticles from tannic acid: synthesis, characterization and evaluation of antileishmanial and cytotoxic activities. An. Acad. Bras. Cienc. **90**, 2679–2689 (2018)
17. Wang, Z., Zhang, Y., Liu, S., Zhang, T.: Preparation of Ag nanoparticles-SnO_2 nanoparticles-reduced graphene oxide hybrids and their application for detection of NO_2 at room temperature. Sens. Actuators, B Chem. **222**, 893–903 (2016)
18. Lin, M., et al.: Growth-dissolution-regrowth transitions of Fe_3O_4 nanoparticles as building blocks for 3D magnetic nanoparticle clusters under hydrothermal conditions. Langmuir **29**, 15433–15441 (2013)
19. Ma, Q., et al.: Fe_3O_4 nanoparticles coated with polyhedral oligomeric silsesquioxanes and β-cyclodextrin for magnetic solid-phase extraction of carbaryl and carbofuran. J. Sep. Sci. **43**, 1514–1522 (2020)
20. de Lima, L.F., et al.: An investigation of the synergistic effect between magnetite nanoparticles and polypyrrole in nanostructured layer-by-layer films. J. Appl. Polym. Sci. **138** (2021)
21. Chen, G.H., et al.: Self-assembly of a Ti4(embonate)6Cage toward silver. Inorg. Chem. **59**, 14861–14865 (2020)
22. Al Sharie, A.H., et al.: Green synthesis of zinc oxide nanoflowers using hypericum triquetrifolium extract: characterization, antibacterial activity and cytotoxicity against lung cancer A549 cells. Appl. Organomet. Chem. **34**, 1–13 (2020)
23. Suchea, M.P., et al.: Article obtaining nanostructured zno onto si coatings for optoelectronic applications via eco-friendly chemical preparation routes. Nanomaterials **11**, (2021)
24. Hassan, K., Hossain, R., Sahajwalla, V.: Novel microrecycled ZnO nanoparticles decorated macroporous 3D graphene hybrid aerogel for efficient detection of NO_2 at room temperature. Sens. Actuators, B Chem. **330**, 129278 (2021)
25. Wang, L., et al.: Fluorescent hybrid nanospheres induced by single-stranded DNA and magnetic carbon quantum dots. New J. Chem. **43**, 4965–4974 (2019)
26. Zhang, C., Miao, P., Sun, M., Yan, M., Liu, H.: Progress in miRNA detection using graphene material-based biosensors. Small **15**, 1–23 (2019)
27. Dong, X.C., et al.: 3D graphene-cobalt oxide electrode for high-performance supercapacitor and enzymeless glucose detection. ACS Nano **6**, 3206–3213 (2012)
28. Nasibulin, A.G., et al.: A novel hybrid carbon material. Nat. Nanotechnol. **2**, 156–161 (2007)
29. Sarkar, S., Dinda, S., Choudhury, P., Kumar Das, P.: Self-assembly of surface functionalized amphiphilic carbon dots: tuning in morphological manifestations. Soft Matter **15**, 2863–2875 (2019)
30. Wareing, T.C., Gentile, P., Phan, A.N.: Biomass-based carbon dots: current development and future perspectives. ACS Nano (2021). https://doi.org/10.1021/acsnano.1c03886
31. Rozhin, P., Charitidis, C., Marchesan, S.: Self-assembling peptides and carbon nanomaterials join forces for innovative biomedical applications. Molecules **26** (2021)
32. Yao, S., Yuan, X., Jiang, L., Xiong, T., Zhang, J.: Recent Progress on Fullerene-Based Materials (2020)

33. Antoku, D., Sugikawa, K., Ikeda, A.: Photodynamic activity of fullerene derivatives solubilized in water by natural-product-based solubilizing agents. Chem. A Eur. J. **25**, 1854–1865 (2019)
34. Shahzad, K., Tahir, M.B., Sagir, M.: Engineering the performance of heterogeneous WO3/fullerene@Ni3B/Ni(OH)2 photocatalysts for hydrogen generation. Int. J. Hydrogen Energy **44**, 21738–21745 (2019)
35. Ciesielski, A., Samorì, P.: Supramolecular approaches to graphene: from self-assembly to molecule-assisted liquid-phase exfoliation. Adv. Mater. **28**, 6030–6051 (2016)
36. Liu, Z., et al.: Ultralight hybrid silica aerogels derived from supramolecular hydrogels self-assembled from insoluble nano building blocks. RSC Adv. **11**, 7331–7337 (2021)
37. Yang, F., et al.: Chirality pure carbon nanotubes: growth, sorting, and characterization. Chem. Rev. **120**, 2693–2758 (2020)
38. Fang, R., et al.: The regulating role of carbon nanotubes and graphene in lithium-ion and lithium-sulfur batteries. Adv. Mater. **31**, 1–22 (2019)
39. Ortiz, E., Gallay, P., Galicia, L., Eguílaz, M., Rivas, G.: Nanoarchitectures based on multi-walled carbon nanotubes non-covalently functionalized with Concanavalin A: a new building-block with supramolecular recognition properties for the development of electrochemical biosensors. Sens. Actuators, B Chem. **292**, 254–262 (2019)
40. Wu, L.L., et al.: Formation of hierarchical NiFe Prussian blue analogues/Prussian blue on nickel foam for superior water oxidation. Appl. Surf. Sci. **567**, 1–8 (2021)
41. Rodrigues, W.V., et al.: Structural reorganization of CuO/Cu2 [Fe(Cn)6] nanocomposite: characterization and electrocatalytic effect for the hydrogen peroxide reduction. An. Acad. Bras. Cienc. **92**, 1–16 (2020)
42. Niu, Q., et al.: Glucose-sensing abilities of mixed-metal (Ni[sbnd]Co) Prussian blue analogs hollow nanocubes. J. Electroanal. Chem. **874**, 114507 (2020)
43. Carvalho, C.L.C., et al.: Development of $Co_3[Co(CN)_6]2/Fe_3O_4$ bifunctional nanocomposite for clinical sensor applications. ACS Appl. Nano Mater. **1**, 4283–4293 (2018)
44. Moraes, R.R., Farias, E.A. de O., Carvalho, C.L.C., Cantanhêde, W., Eiras, C.: Development of cashew gum-based bionanocomposite as a platform for electrochemical trials. Int. J. Biol. Macromol. **153**, 118–127 (2020)
45. Guo, J., Suma, T., Richardson, J.J., Ejima, H.: Modular assembly of biomaterials using polyphenols as building blocks. ACS Biomater. Sci. Eng. **5**, 5578–5596 (2019)
46. Mathivanan, N., et al.: Hydrogen-bonded multilayer thin films and capsules based on poly(2-n-propyl-2-oxazoline) and tannic acid: investigation on intermolecular forces, stability, and permeability. Langmuir **35**, 14712–14724 (2019)
47. Abe, I., et al.: A trigonal molecular assembly system with the dual light-driven functions of phase transition and fluorescence switching. J. Mater. Chem. C **7**, 2276–2282 (2019)
48. Abe, I., Han, M.: Green-light-induced melting of self-assembled azobenzene nano/microstructures. New J. Chem. **43**, 19014–19019 (2019)
49. Yin, H., Wang, H., Li, Z., Shu, D., Guo, P.: RNA micelles for the systemic delivery of anti-miRNA for cancer targeting and inhibition without ligand. ACS Nano **13**, 706–717 (2019)
50. Liu, R., Hudalla, G.A.: Using self-assembling peptides to integrate biomolecules into functional supramolecular biomaterials. Molecules **24**, 15–21 (2019)
51. Moon, H., Lee, C., Lee, W., Kim, J., Chae, H.: Stability of quantum dots, quantum dot films, and quantum dot light-emitting diodes for display applications. Adv. Mater. **31**, 1–14 (2019)
52. Hasan, S., et al.: Thermodynamic modelling of InAs/InP(0 0 1) growth towards quantum dots formation by metalorganic vapor phase epitaxy. J. Cryst. Growth **509**, 133–140 (2019)
53. Sun, R., et al.: Dual quantum dots decorated TiO_2 nanorod arrays for efficient CO_2 reduction. J. Catal. **378**, 192–200 (2019)
54. Sowa, A., Voskuhl, J.: Host-guest complexes–boosting the performance of photosensitizers. Int. J. Pharm. **586**, 119595 (2020)
55. Khurana, R., et al.: Supramolecular nanorods of (N-methylpyridyl) porphyrin with captisol: effective photosensitizer for anti-bacterial and anti-tumor activities. Front. Chem. **7**, 1–11 (2019)

56. Shee, N.K., Kim, M.K., Kim, H.J.: Fluorescent chemosensing for aromatic compounds by a supramolecular complex composed of tin(iv) porphyrin, viologen, and cucurbit[8]uril. Chem. Commun. **55**, 10575–10578 (2019)
57. Yao, T.T., et al.: A photodynamic antibacterial spray-coating based on the host-guest immobilization of the photosensitizer methylene blue. J. Mater. Chem. B **7**, 5089–5095 (2019)
58. Tang, Y., et al.: Redox-responsive poly(ionic liquid) microgels explored as the building blocks for supramolecular assembly. Polymer (Guildf) **220**, 123575 (2021)
59. Ahmad, R., et al.: Deposition of nanomaterials: a crucial step in biosensor fabrication. Mater. Today Commun. **17**, 289–321 (2018)
60. Cui, K., Dorner, I., Mertens, S.F.L.: Interfacial supramolecular electrochemistry. Curr. Opin. Electrochem. **8**, 156–163 (2018)
61. Hein, R., Beer, P.D., Davis, J.J.: Electrochemical anion sensing: supramolecular approaches. Chem. Rev. **120**, 1888–1935 (2020)
62. Ran, X., et al.: Water-soluble pillar[6]arene functionalized nitrogen-doped carbon quantum dots with excellent supramolecular recognition capability and superior electrochemical sensing performance towards TNT. Sens. Actuators, B Chem. **257**, 362–371 (2018)
63. Boumya, W., Taoufik, N., Achak, M., Barka, N.: Chemically modified carbon-based electrodes for the determination of paracetamol in drugs and biological samples. J. Pharm. Anal. **11**, 138–154 (2021)
64. Watson, S., Nie, M., Wang, L., Stokes, K.: Challenges and developments of self-assembled monolayers and polymer brushes as a green lubrication solution for tribological applications. RSC Adv. **5**, 89698–89730 (2015)
65. Alotaibi, H.F., Al Thaher, Y., Perni, S., Prokopovich, P.: Role of processing parameters on surface and wetting properties controlling the behaviour of layer-by-layer coated nanoparticles. Curr. Opin. Colloid Interface Sci. **36**, 130–142 (2018)
66. Li, X., et al.: Conducting polymers in environmental analysis. TrAC Trends Anal. Chem. **39**, 163–179 (2012)
67. Rizzo, H.F., Bidwell, L.R.: Formation and structure of SiB4. J. Am. Ceram. Soc. **43**, 550–552 (1960)
68. Singh, M., Kaur, N., Comini, E.: The role of self-assembled monolayers in electronic devices. J. Mater. Chem. C **8**, 3938–3955 (2020)
69. Kong, G.D., et al.: Mixed molecular electronics: tunneling behaviors and applications of mixed self-assembled monolayers. Adv. Electron. Mater. **6**, 1–19 (2020)
70. Liu, Y., Qiu, X., Soni, S., Chiechi, R.C.: Charge transport through molecular ensembles: recent progress in molecular electronics. Chem. Phys. Rev. **2**, 021303 (2021)
71. Lin, Y., et al.: Self-assembled monolayer enables hole transport layer-free organic solar cells with 18% efficiency and improved operational stability. ACS Energy Lett. **5**, 2935–2944 (2020)
72. Kong, G.D., et al.: Interstitially mixed self-assembled monolayers enhance electrical stability of molecular junctions. Nano Lett. **21**, 3162–3169 (2021)
73. Qi, L., Tian, H., Shao, H., Yu, H.Z.: Host-guest interaction at molecular interfaces: Cucurbit[7]uril as a sensitive probe of structural heterogeneity in ferrocenyl self-assembled monolayers on gold. J. Phys. Chem. C **122**, 15986–15995 (2018)
74. Li, Y., Li, Y., Zhang, Y., Song, Y., Jiang, Y.: A rapid and sensitive electrochemical sensor for hydroxyl free radicals based on self-assembled monolayers of carboxyl functionalized graphene. J. Solid State Electrochem. **23**, 187–194 (2019)
75. Xu, X., et al.: Structural changes of mercaptohexanol self-assembled monolayers on gold and their influence on impedimetric aptamer sensors. Anal. Chem. (2019). https://doi.org/10.1021/acs.analchem.9b03946
76. Yılmaz Aykut, D., Yolaçan, Ö., Deligöz, H.: pH stimuli drug loading/release platforms from LbL single/blend films: QCM-D and in-vitro studies. Colloids Surfaces A Physicochem. Eng. Asp. **602** (2020)
77. Ariga, K., Ahn, E., Park, M., Kim, B.S.: Layer-by-layer assembly: recent progress from layered assemblies to layered nanoarchitectonics. Chem. An Asian J. **14**, 2553–2566 (2019)

78. Zhang, X., Chen, H., Zhang, H.: Layer-by-layer assembly: from conventional to unconventional methods. Chem. Commun. 1395–1405 (2007). https://doi.org/10.1039/b615590a
79. Li, Y., Wang, X., Sun, J.: Layer-by-layer assembly for rapid fabrication of thick polymeric films. Chem. Soc. Rev. **41**, 5998–6009 (2012)
80. Wang, S., et al.: Graphene quantum dot-assisted preparation of water-borne reduced graphene oxide/polyaniline: from composite powder to layer-by-layer self-assembly film and performance enhancement. Electrochim. Acta **295**, 29–38 (2019)
81. Wu, S., et al.: Layer-by-layer self-assembly film of PEI-reduced graphene oxide composites and cholesterol oxidase for ultrasensitive cholesterol biosensing. Sens. Actuators, B Chem. **298**, 126856 (2019)
82. Correia, A.R., Sampaio, I., Comparetti, E.J., Vieira, N.C.S., Zucolotto, V.: Optimized PAH/Folic acid layer-by-layer films as an electrochemical biosensor for the detection of folate receptors. Bioelectrochemistry **137**, 107685 (2021)
83. Guzmán, E., Rubio, R.G., Ortega, F.: A closer physico-chemical look to the layer-by-layer electrostatic self-assembly of polyelectrolyte multilayers. Adv. Colloid Interface Sci. **282**, 102197 (2020)
84. Gu, M., Kim, B.S.: Electrochemistry of multilayer electrodes: from the basics to energy applications. Acc. Chem. Res. **54**, 57–69 (2021)
85. Wang, Z., et al. Layer-by-layer self-assembled nanostructured electrodes for lithium-ion batteries. Small **17** (2021)
86. Wang, S., et al.: High-performance layer-by-layer self-assembly PANI/GQD-rGO/CFC electrodes for a flexible solid-state supercapacitor by a facile spraying technique. ACS Appl. Energy Mater. **2**, 1077–1085 (2019)
87. Pu, J., et al.: Electrodeposition technologies for Li-based batteries: new frontiers of energy storage. Adv. Mater. **32**, 1–28 (2020)
88. Palomar-Pardavé, M., et al.: Supramolecular interaction of dopamine with β-cyclodextrin: an experimental and theoretical electrochemical study. J. Electroanal. Chem. **717–718**, 103–109 (2014)
89. Healy, B., Yu, T., da Silva Alves, D.C., Okeke, C., Breslin, C.B.: Cyclodextrins as supramolecular recognition systems: applications in the fabrication of electrochemical sensors. Materials (Basel) **14** (2021)
90. Unger, C., Lieberzeit, P.A.: Molecularly imprinted thin film surfaces in sensing: chances and challenges. React. Funct. Polym. **161**, 104855 (2021)
91. Uppachai, P., Srijaranai, S., Poosittisak, S., Isa, I.M., Mukdasai, S.: Supramolecular electrochemical sensor for dopamine detection based on self-assembled mixed surfactants on gold nanoparticles deposited graphene oxide. Molecules **25** (2020)
92. Esmaeily, Z., Madrakian, T., Afkhami, A., Ghoorchian, A., Ghasemzadeh-Mohammadi, V.: Electropolymerization as an electrochemical preconcentration approach for the determination of melamine in milk samples. Electrochim. Acta **390**, 138897 (2021)
93. Mangadlao, J.D., et al.: Grafted carbazole-assisted electrodeposition of graphene oxide. ACS Appl. Mater. Interfaces **7**, 10266–10274 (2015)
94. Ohtani, Y., Kumano, K., Saneshige, M., Takami, K., Hoshi, H.: Effect of electropolymerization duration on the structure and performance of polypyrrole/graphene nanoplatelet counter electrode for dye-sensitized solar cells. J. Solid State Electrochem. **25**, 2107–2113 (2021)
95. Yang, J., et al. Electropolymerization process dependent poly(1,4-di(2-thienyl)benzene) based full spectrum activated photocathodes for efficient photoelectrochemical hydrogen evolution. J. Electroanal. Chem. **903**, 115712 (2021)
96. Zhang, D., Zhang, J., Pan, M., Wang, Y., Sun, T.: Necklace-like C-ZIF-8@MWCNTs fabricated by electrochemical deposition towards enhanced supercapacitor. J. Alloys Compd. **853**, 157368 (2021)
97. Abbasi Kajani, A., Haghjooy Javanmard, S., Asadnia, M., Razmjou, A.: Recent advances in nanomaterials development for nanomedicine and cancer. ACS Appl. Bio Mater. (2021). https://doi.org/10.1021/acsabm.1c00591

98. Hu, Q., Li, H., Wang, L., Gu, H., Fan, C.: DNA nanotechnology-enabled drug delivery systems. Chem. Rev. **119**, 6459–6506 (2019)
99. Baig, N., Sajid, M., Saleh, T.A.: Recent trends in nanomaterial-modified electrodes for electroanalytical applications. TrAC Trends Anal. Chem. **111**, 47–61 (2019)
100. Da Silva, E.T.S.G., et al.: Electrochemical biosensors in point-of-care devices: recent advances and future trends. ChemElectroChem **4**, 778–794 (2017)
101. Roushani, M., Rahmati, Z., Farokhi, S., Hoseini, S.J., Fath, R.H.: The development of an electrochemical nanoaptasensor to sensing chloramphenicol using a nanocomposite consisting of graphene oxide functionalized with (3-Aminopropyl) triethoxysilane and silver nanoparticles. Mater. Sci. Eng. C **108**, 110388 (2020)
102. Gayda, G.Z., et al.: Metallic nanoparticles obtained via 'green' synthesis as a platform for biosensor construction. Appl. Sci. **9**, (2019)
103. Avelino, K.Y.P.S., Oliveira, L.S., Lucena-Silva, N., Andrade, C.A.S., Oliveira, M.D.L.: Flexible sensor based on conducting polymer and gold nanoparticles for electrochemical screening of HPV families in cervical specimens. Talanta **226** (2021)
104. Avelino, K.Y.P.S., et al. Nanostructured sensor platform based on organic polymer conjugated to metallic nanoparticle for the impedimetric detection of SARS-CoV-2 at various stages of viral infection. J. Pharm. Biomed. Anal. **206**, 114392 (2021)
105. Fatema, K.N., Oh, W.C.: A comparative electrochemical study of non-enzymatic glucose, ascorbic acid, and albumin detection by using a ternary mesoporous metal oxide (ZrO_2, SiO_2 and In_2O_3) modified graphene composite based biosensor. RSC Adv. **11**, 4256–4269 (2021)
106. Hashemi, S.A., et al.: Coupled graphene oxide with hybrid metallic nanoparticles as potential electrochemical biosensors for precise detection of ascorbic acid within blood. Anal. Chim. Acta **1107**, 183–192 (2020)
107. Xia, J., Zou, B., Liu, F., Wang, P., Yan, Y.: Sensitive glucose biosensor based on cyclodextrin modified carbon nanotubes for detecting glucose in honey. J. Food Compos. Anal. **105**, 104221 (2022)
108. Zamiri, G., et al.: Ternary nanocomposite cathodes based on 3D graphene-Ag nanoparticle-polyaniline for hybrid electrochemical energy device. Synth. Met. **282**, 116932 (2021)
109. Cao, R., Díaz-García, A.M., Cao, R.: Coordination compounds built on metal surfaces. Coord. Chem. Rev. **253**, 1262–1275 (2009)
110. Carvalho, C.L.C., et al.: Effect of Ibuprofen on the electrochemical properties of Prussian blue/single-walled carbon nanotubes nanocomposite modified electrode. Surfaces Interfaces **25** (2021)

In Situ and Operando Techniques in Bioelectrochemistry

João C. Perbone de Souza

1 Introduction

Bioelectrochemistry is a field that use the electrochemistry to investigate biological systems and brings together scientists from variety of areas including chemistry, biology, and physics. The electrochemistry can provide information about thermodynamics and kinetics of the electron transfer processes on biological systems; thus, it is possible to know how the biological cells capture energy and the mechanistic pathways involved in the biological reactions. However, the bioelectrochemical approach provide no information about the changes on the molecular structures. To solve this limitation, the merge of the electrochemical with another technique (spectroscopic, spectrometric, and microscopic) appears as a solution [1–6]. These techniques are known as in situ or operando characterization, which can provide in-depth insights into the biological reactions, thereby strongly contributing to obtaining hints, such as the nature of the enzymatic active site and reaction mechanisms. By definition, operando is a branch of in situ techniques that monitors a working catalyst during genuine reaction condition with simultaneous measurement of catalytic activity and selectivity [1].

Figure 1 presents a scheme of some kinds of spectroscopies that can be combined with bioelectrochemical techniques. To exemplify, there is presented a multicopper oxidase as generic investigated biomolecule, which has two redox centers. This kind of biomolecule has an internal process known as the intramolecular electron transfer (ET) that can be measured indirectly by spectroscopy [7], by the quantification of substrate/products. There is also the interfacial ET, which occurs between enzyme and electrode, this process can be measured by using a potentiostat/galvanostat.

J. C. P. de Souza (✉)
Goiano Federal Institute of Education, Science and Technology, Campus Rio Verde, Rio Verde, Goiás 75901-970, Brazil
e-mail: joao.perbone@ifgoiano.edu.br

F. N. Crespilho (ed.), *Advances in Bioelectrochemistry Volume 1*,
https://doi.org/10.1007/978-3-030-94988-4_7

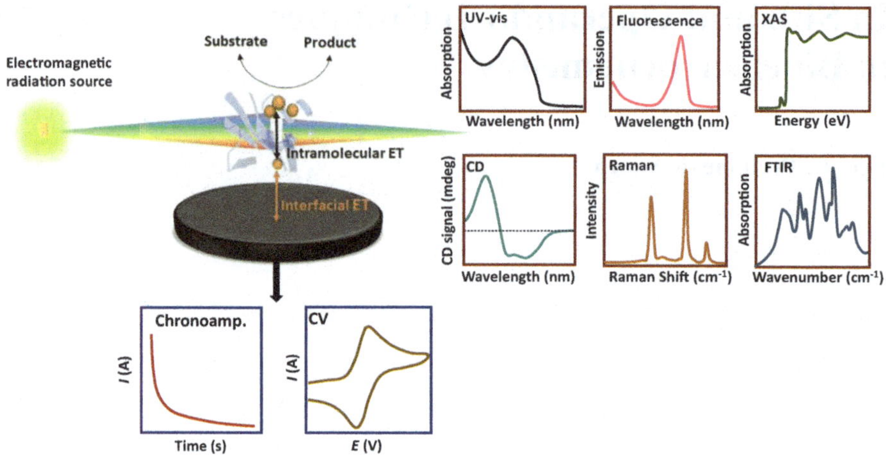

Fig. 1 Schematic representation of an in situ or operando spectroscopy in a bioelectrochemical systems. The interaction of electromagnetic source with biomolecule onto electrodic surface provides spectroscopic data that depends on range of electromagnetic spectrum is the incident radiation. The electrochemical connection can provide the potential/current control for the bioelectrode and give information of the electron transfer of adsorbed biomolecules on electrode

The information obtained from spectroscopy depends on the region of electromagnetic radiation is used light source. The quantum changes (electronic, vibrational, and rotational) in molecules and atoms depends on the incident electromagnetic radiation, since the energy of ranges of electromagnetic spectra are different. The spectroscopy can be applied in different ways, for instance, the electronic absorption techniques allow the use the Lambert–Beer principle to quantify chemical species, while the vibrational spectroscopy provides information about ligands at active site of an enzyme. The versatility of spectroscopy gives a wide range to analyze several properties, such as kinetics, thermodynamics, and binding parameters. Likewise, the bioelectrochemistry is useful tool to assess physicochemical properties more related to interface process.

The mass spectrometry and microscopy can be also merged with bioelectrochemistry and can brings important information on enzymatic kinetics and biomolecules organization in a cell, respectively. These kind of operando and in situ techniques will be detailed in this chapter.

2 Restraints Involved in Combining Bioelectrochemistry with Analytical Techniques

The electrical communication between surface and biomolecule plays an important role for bioelectrochemical measurements. In order to establish an effective electrical relay from biomolecule to electrode and vice versa, several immobilization methods have been developed [8]. However, some strategies are not suitable for operando and in situ bioelectrochemistry, due to the use of some chemical compounds that interferes in spectroscopy signal and not provide an adequate molecular orientation on electrode. Other restraint arises from the low coverage of enzymes on an electrode, for example, an enzyme molecule of 5 nm diameter and densely packed monolayer on a planar electrode surface would have maximum coverage of *ca* 8 pmol cm^{-2} [6].

The biological activity of biomolecules depends on the spatial conformation, in specially, for proteins. Some physical and chemical factors can change the biomolecular structure, such as the heat caused by intensity of light source [9]. Moreover, the energy of electromagnetic source, which is higher for lower wavelengths values, can modify the structure of biomolecules by ionization or cleaving some chemical bond [10]. Thereby the cell design plays a key role for a suitable measure of bioelectrochemical systems by using in situ or operando techniques.

Cell design depends on the mode of spectroscopy sampling that is used. The transmission mode is applicable for (semi-)transparent working electrode, which is used in ultraviolet–visible (UV–Vis), infrared (IR), X-ray absorption (XAS) or MCD spectroscopy [6, 11, 12]. Thus, the bioelectrochemical cell design must allow the passage of radiation through the working bioelectrode. The external reflection and the attenuated total reflectance modes that is more used in vibrational spectroscopies, must have a bioelectrochemical cell that permits the reflection of light that leaves the surface of bioelectrode.

Electrode materials play an important role [13, 14], since they can have active signals for some kinds of spectroscopies and the spectroscopy signal of electrode can overlap the signal from biomolecules. Metallic electrodes must be avoided in magnetic spectroscopies due to application of strong magnetic field [15]. In addition, the material influences the molecular orientation of biomolecules on surface [16]. Thus, the observation of some bioelectrochemical systems with accuracy depends on the knowledge of this properties.

3 In Situ and Operando Spectroscopies in Bioelectrochemistry

This topic is divided in the spectroscopic analytical tools that have been merged with biolectrochemistry to provide in situ and operando techniques. The quantum changes depend on wavelength of electromagnetic radiation source and the detector, there is possible to observe electronic, vibrational, and rotational transitions. Thus, the subtopics are organized by describing firstly the spectroscopies that use more energetic radiation sources and lastly the less energetic spectroscopies. The next topic describes operando mass spectrometry and microscopy, which are non-spectroscopic techniques.

3.1 *Operando and In Situ Electronic Spectroscopies in Biolectrochemistry*

3.1.1 X-ray Absorption Spsectroscopy

The X-ray absorption spectroscopy (XAS) can be used as a direct probe of the oxidation state and of the geometric and electronic structure of metallic sites in proteins [17], such as copper [18], nickel [19], and iron [20]. The resulting XAS spectra is divided into two regions of interest. The first region is the X-ray absorption near edge structure (XANES), which is in the spectral range of the 50 eV around the absorption edge and provides information about the electronic structure, density of unoccupied states, and bonding geometry around the absorbing atom (element-specific). The second region is the extended X-ray absorption fine structure (EXAFS) that is defined by the modulation of the absorption coefficient up to 1000–1500 eV after the absorption edge, which can be used to probe the local structure [21]. Thereby, this kind of spectroscopy, specially XANES, is very suitable to investigated redox centers in metalloproteins.

The XAS technique when used as a standalone technique in non-electrodic ex situ redox experiments provide important information on the electronic behavior of some metals and supplies insights into the biological role of such metal-containing structures [22–24]. Nevertheless, the nature of ligand structure around coordination centers that has great importance in biological activity and biomimetic systems is not evaluated just by using XAS, this information can only be obtained by combining in situ XAS with electrochemistry. For metalloproteins, the modulation of the oxidation state of metallic sites can be provided by bioelectrochemistry techniques, such as protein film voltammetry (PFV) [25]. The redox state can be changed under precise potential control; thus, it is possible to assess the changes on electronic structure by using XAS.

In this sense, operando XAS was used to investigate the pathway adopted by bilirubin oxidase from *Myrothecium verrucaria* (*Mv*BOD) that leads to the reduction of O_2 to H_2O [26]. The *Mv*BOD is a multicopper enzyme, thereby the operando XAS assessed the redox activity of the Cu ions (T1, T2/T3) in effecting the electron transfer from the substrate/electrode surface to the molecular oxygen under catalytic and non-catalytic conditions. Figure 2a–d shows details of the crystal structure of *Mv*BOD (PDB: 2XLL) zoomed in at the four Cu ions and the schematic illustration of operando XAS setup by using synchrotron radiation.

Figure 2e shows the X-ray absorption spectra of *Mv*BOD recorded by applying different potentials, which illustrates the evolution of a redox reaction. There was observed the reduction of Cu^{2+}, as presented by the attenuation of the signal at 8997 eV and the concomitant increase of the signal at 8983 eV due to the presence of Cu^{+} in the enzymatic structure. When compared to the case of absence of O_2 in the bioelectrochemical system, the bonding of O_2 to the trinuclear cluster provided a 150 mV increase of the onset potential in the potentiometric titration curve (Fig. 2f) in the presence of Cu^{+}. This phenomenon suggests the occurrence of fast electron

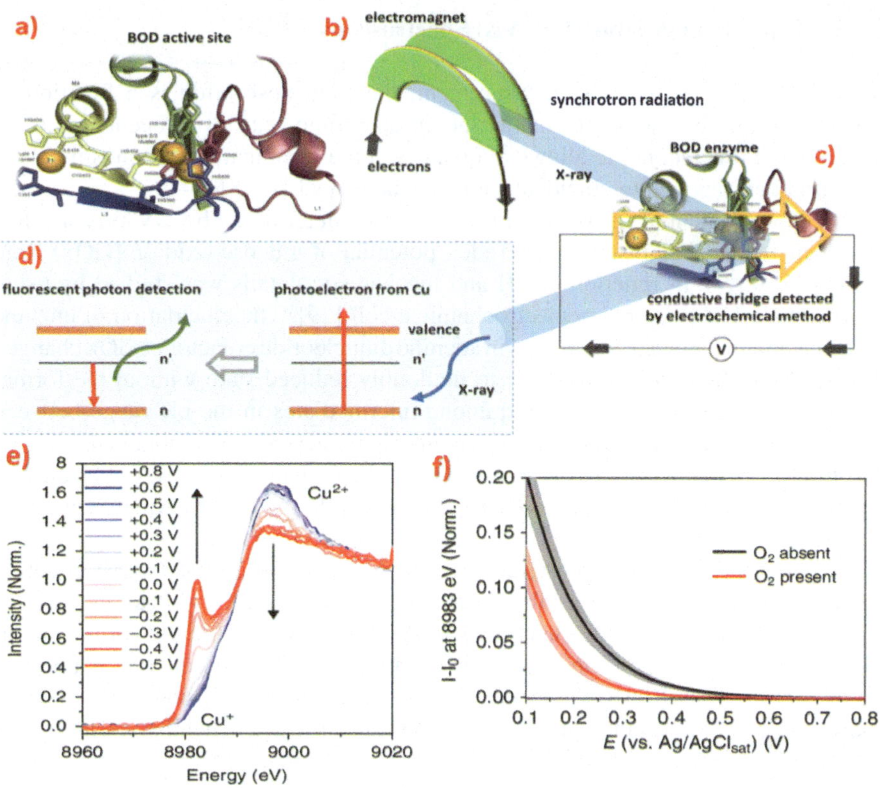

Fig. 2 Assessment of ET reactions and catalysis in multicopper oxidases using operando XAS. **a** Representation of the crystal structure of *Mv*BOD (PDB: 2XLL) zoomed in at the four Cu ions. **b** Schematics of synchrotron radiation generation. The XAFS2 beamline of the Brazilian Synchrotron Light Laboratory (LNLS), a national institution responsible for operating the only synchrotron light source in Latin America, was used. **c** Principle of XAS spectroelectrochemical analysis. **d** Simplified scheme showing the fundamental processes required to obtain in situ XANES spectra during bioelectrochemical characterization, with X-ray photoabsorption of core-level electrons followed by photoelectron emission and fluorescence. **e** K-edge spectra of *Mv*BOD under different electrochemical potentials. **f** Potentiometric curves of Cu^+ signal intensity under each atmospheric condition highlighting the rise of this signal at different overpotentials. Reprinted from Ref. [26] with permission from the Nature Publishing Group

transfer from T1 to the trinuclear cluster and the ability of the metal cofactor to act as an electronic bridge connecting the electron donor (organic substrate or electrode surface) and the electron acceptor (O_2) in two distant parts of the enzyme for a faster and more energetically favorable electron transfer [26].

3.1.2 Ultraviolet–Visible (UV–Vis) Spectroscopy

In an UV–Vis spectroscopic measurement, the light absorption is a function of wavelength that provides information about electronic transitions occurring in the material and this technique allows to quantify some chemical compounds by using Beer-Lambert law and to elucidate some mechanisms [27, 28].

The redox potential of oxidoreductases can be determined by UV–Vis spectro-electrochemistry. For instance, the redox potential of glucose oxidase (GOx) from *Aspergillus niger* in function of pH and mechanistic details was studied by using a homemade UV–Vis spectroelectrochemical cell [29]. The elucidation of enzyme redox chemistry showed that the flavin adenine dinucleotide cofactor of GOx changed directly from the oxidized quinone to its doubly reduced state without the formation of stable singly reduced semiquinone intermediates in the pH range between 4.5 and 8.5. Other enzymatic thermodynamics parameters can be obtained by using spectroelectrochemical UV–Vis. For instance, the enthalpy, entropy, Gibbs energy, and reduction potential of *Schizosaccharomyces pombe (Sp)* ferredoxin was obtained with in situ UV–Vis [30].

In bioelectrochemistry, most redox proteins do not obey the Nernst equation because of the large steric inhibition to electron transfer, as well as the way of enzymatic adsorption on the electrode surface. For the calculation of formal potential (E°) and the electron transfer number (n) from the Nernst equation is possible by monitoring the absorbance changes in situ as a function of potential [31], this shows the great importance of in situ UV–Vis spectroelectrochemistry. The Nernst equation is

$$E = E^{\circ} + \frac{RT}{nF} \ln \frac{[O]}{[R]}$$

where E° is the formal potential of the redox couple O/R, R is the molar gas constant (8.314 J mol^{-1} K^{-1}), and F is Faraday's constant (96,485 C mol^{-1}). $[O]$ and $[R]$ are the concentrations of the oxidized and reduced species, respectively. From the plot of E *versus* ln ($[O]/[R]$) is possible to determine the formal potential E° and the electron transfer number n. The ratio $[O]/[R]$ is determined by using the Lambert–Beer law:

$$\frac{[O]}{[R]} = \frac{\left(\frac{A-A_R}{\Delta\varepsilon\, b}\right)}{\left(\frac{A_O-A}{\Delta\varepsilon\, b}\right)} = \frac{A - A_R}{A_O - A}$$

where A_R is the absorbance of the reduced form, A_O is the absorbance of the oxidized form, A is the absorbance of an intermediate mixture, $\Delta\varepsilon$ is the molar absorptivity difference of the oxidized and reduced form measured at the same wavelength, and b is the optical path length [31]. If the experiments will be carried out with temperature changes, there is possible to calculate entropy and enthalpy changes.

3.1.3 Fluorescence Spectroscopy

Fluorescence spectroscopy is based on luminescence properties of chemical species, which is featured by substances that emits light provided from electronically excited states [32]. This technique is highly sensitive, and the fluorescent probes can be used to establish the degree of polarity of a particular region of a protein; measure distances between groups in a protein; determine the extent of flexibility of a protein [33, 34] and nucleic acids [35, 36]; and measure the rate of very rapid conformational transitions.

In fluorescence spectroelectrochemistry, the electromagnetic phenomenon of Föster or fluorescence resonance energy transfer (FRET) is the most used to elucidate mechanisms involving biomolecules [37]. FRET is the radiationless transfer of energy between two light-sensitive molecules (donor and acepptor), this physical process depends on effective spectral overlap and appropriate fluorophore dipole alignment. The phenomenon is characterized by a decrease in donor emission with simultaneous increasing in acceptor emission, is a highly distance-sensitive measurement, thus can be used to measure protein–protein interactions and conformational fluctuations within suitably labeled individual molecules. The fluorescence spectroelectrochemistry was used to study the electrochemical behavior of azurin, which is a metalloprotein with a redox-active Cu ion center that has a strong absorption band around 550–600 nm in the oxidized state that disappears in the reduced state [38–40]. This transition corresponds to π–π^* transition centered at the copper site and involving mainly the dx^2–y^2 orbital and a 3p orbital on the Cys112 ligand sulfur, the copper site can be reduced to a d^{10} electronic configuration and then the absorption disappears (inset Fig. 3a) [41]. These studies monitored fluorophore labelled azurin immobilized on self-assembled monolayer (SAM) coated with gold thin-film electrodes, as schematically represent in Fig. 3a. An artistic representation of the fluorescence electrochemistry assembly combined with confocal fluorescence microscopy is represented in Fig. 3b [42].

The in situ spectroeletrochemical fluorescence has been used to study other biomolecules, such as nucleic acids [43–46] and coenzymes [47, 48]. For instance, the SAM containing DNA, Aib peptide and alkylthiol was studied in function of the electrical potential of the interface revealing how the organization of these SAM depend on the surface crystallographic orientation [46]. By fluorescence imaging of mercaptohexanol with single strain DNA SAM on a single crystal gold bead electrode, it was possible to observe a clear and substantial influence of the surface crystallography on the competitive displacement of mercaptohexanol by single strain DNA. In the case of coenzymes that have tricyclic heteronuclear organic ring in their structures, the fluorescence spectroelectrochemistry can be effectively applied to explore the redox reactions of these molecules, then gaining insights on biological mechanism [47].

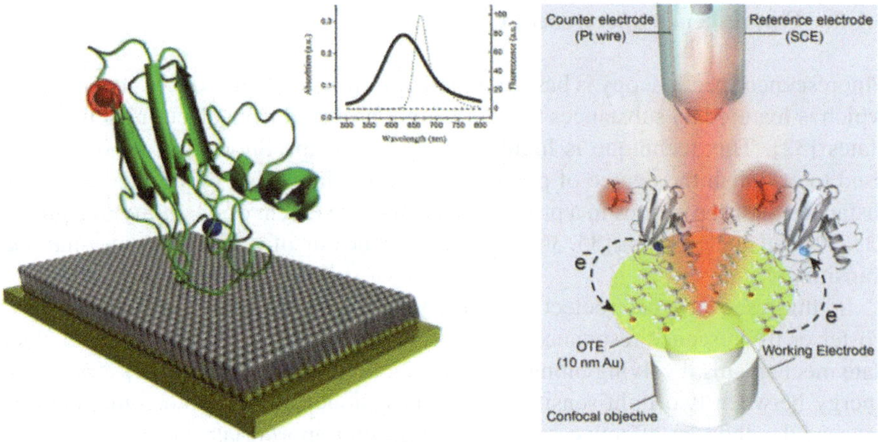

Fig. 3 **a** Cartoon representation of N-terminus labelled azurin on SAM modified gold with an appended fluorophore depicted as a red sphere (the copper shown as a blue sphere). Inset: optical spectrum overlaps for the azurin Cu^{2+} Cy5 FRET pair. The Cy5 emission spectrum (dotted line) was determined at excitation wavelength of 600 nm. The UV–Vis upon reduction of the metal redox center (dashed line), eliminating the spectral overlap between the fluorophore and redox center and turning the pathway to non-radiative energy loss from the fluorophore "off" (in effect, turning emission from the fluorophore "on"). **b** An artistic representation of the fluorescence electrochemistry assembly combined with confocal fluorescence microscopy. A reference electrode (saturated calomel electrode, SCE) and counter electrode (Pt wire) were inserted in the sample solution which was contained in a sealed holder with a semi-transparent, gold-coated glass cover slip at the base which functioned as the optically transparent working electrode. Azurin labeled with Cy5 is adsorbed on the SAM of octanethiol which was deposited on the gold film. Fluorescence of the dye label is quenched when the protein is oxidized because of FRET from the attached fluorophore to the metal-cofactor (dark blue sphere). In the reduced form of the protein (light blue sphere), quenching does not occur, and emission of the fluorophore is uninhibited. Reproduced from Refs. [41, 42] with permission

3.1.4 Circular Dichroism

Circular dichroism (CD) is a measure of the differential absorbance between left- and right-circularly polarized light. CD signals arise from absorption of radiation, and thus spectral bands are easily assigned to distinct structural features of a molecule. An advantage of the CD technique in studies of biomolecules is that complementary structural information can be obtained from some spectral regions, such as peptide bond (below 240 nm), aromatic aminoacid side chains (260–320 nm), disulphide bond (~260 nm), cofactors (distinguished regions) [49], quadruplex DNA (220–300 nm) [50]. By using CD, it is possible to obtain different types of information about secondary structure composition from the peptide bond region, tertiary structure fingerprint from aromatic aminoacid region, and the integrity of cofactor binding sites [49]. A similar technique based on the same principle is the magnetic circular dichroism (MCD), which the sample is exposed in a strong magnetic field parallel to the direction of radiation propagation and is also very useful to characterize biomolecules [51].

Despite the great importance of the role of chirality and biomolecules folding, the CD is still one of the techniques less frequently combined with the electrochemical methods [52]. Nonetheless, some examples of in situ CD applied to bioelectrochemistry is given here. The in situ CD spectroelectrochemistry was used to monitor changes on secondary and tertiary or quaternary structure of hemoglobin [53]. There was observed quaternary structural changes associated with the relaxed to tense transition, i.e., the change from oxidized to deoxidized hemoglobin structure. The electroreduction process of hemoglobin by electron transfer/chemical reaction mechanism is followed by the relaxed to tense transition. Other study evaluated the cyclic denaturation and renaturation of double-stranded DNA by switching redox-state of daunomycin that is a DNA intercalator [54]. The cycles of electrochemically controlled DNA denaturation and renaturation by switching of the daunomycin redox state, were confirmed using in situ UV–Vis and CD spectroelectrochemistry. A clear switching of the CD signal (~275 nm) was observed for DNA in the presence of daunomycin that corroborated with in situ UV–Vis data. The great importance of this study was demonstrated the proof-of-principle of a new electrochemical method to control reversible DNA hybridization as an alternative to heating and cooling of reaction solutions. Another report used in situ MCD for the evaluation of the hemeproteins with electrochemical potential control, this technique brings information of spin state, then the iron center in hemeproteins can be well characterized [55].

3.2 Vibrational Spectroelectrochemistry in Bioelectrochemistry

3.2.1 Infrared Spectroscopy

Vibrational spectroscopy analyzes the periodic oscillations of atoms within a molecule. These oscillations occur in a precisely defined manner [56], i. e., energy of a vibration is quantized in discrete levels. The vibrational spectroscopy has some advantages in characterization of biomolecules: (1) it can contribute to the elucidation of details in the molecular structures and intermolecular; (2) it is in principle not restricted by the size of the sample, thus can provide valuable information for small biomolecules in addition to complex biological systems; (3) vibrational spectroscopic methods are applicable regardless of the aggregation state of the biomolecule [56]. The two main techniques used to obtain vibrational spectra, infrared (IR) and Raman spectroscopy, which are based on different physical mechanism. Noteworthy, the most common form of IR spectroscopy is the Fourier transform infrared (FTIR). Here, the IR spectroscopy coupled with electrochemistry to investigate biomolecules will be detailed.

The in situ and operando IR spectroscopy in bioelectrochemistry is used to investigate redox process of metalloenzymes [57–59], proteins at electrochemical interfaces [60], biofilms [61], and conformational changes of proteins [62–64]. Then, the IR bio/spectroelectrochemistry has shown its great importance in investigation of biological systems. For example, the guiding principles of enzymatic catalysis can be clarified by using protein film IR electrochemistry (PFIRE), this technique reveals the steady-state distribution of intermediates during catalysis of [NiFe] hydrogenases [57] and potential-dependence of CO binding to the FeMo-cofactor of a nitrogenase [65]. Figure 4a shows the structure of [NiFe] hydrogenase I from *E. coli* and the FeS clusters which are electronically linked with the [NiFe] active site that allows the fast electron transfer during the catalysis. Figure 4b presents the currently accepted catalytic cycle for [NiFe] hydrogenases, which possess native CO and CN^- ligands coordinated to the iron atom of their catalytic site. These ligands function as excellent IR absorbers and reporters of electronic and coordination changes at the active site. The iron atom at active site remains the Fe^{2+} during the reactions, while the redox state of Ni changes (Ni_a–R/Ni^{2+}; Ni_a–C/Ni^{3+}; Ni_a–L/Ni^{1+}; Ni_a–SI/Ni^{2+}; Ni–B/Ni^{3+}; subscript “a” means aerobic). There is substantial delocalization as indicated by changes in CO and CN^- wavenumber positions (νCO and νCN) in IR spectra, since they are highly sensitive to electron density on Fe. The Fig. 4c shows a chronoamperogram that is observed a positive catalytic current at −74 and +356 mV, which is related with enzymatic H_2 oxidation channels electrons to the electrode. The current is very stable during the measurement time at −74 mV, but it decays slowly at +356 mV due to well-established oxidative inactivation. Figure 4d presents the IR spectra at these potentials, there is possible to observe changes on the CO bands that is strongly related with the population of intermediates of the catalytic site [66]. Figure 4e shows a spectroscopic data that is turnover (H_2-saturated) *minus* nonturnover (Ar-saturated) difference spectra to reflect changes in speciation in response to catalysis. The positive peaks are provided by species that are populated in response to turnover, the species population is highlighted at Fig. 4f [57].

3.2.2 Raman Spectroscopy

Raman spectroscopy is also a vibrational spectroscopy complementary to IR spectroscopy, which is based on measurement of scattering of radiation by matter, while IR spectroscopy measures the excitation of molecular vibrations by direct absorption. The electromagnetic radiation source used in Raman spectroscopy is a monochromatic light (laser) with a very intense beam. The inelastic scattering of the incident radiation is called Raman scattering, which the radiation leaves the sample with lower or higher frequency, then brings information of the molecular structure and composition of matter [67]. The Raman spectroscopy has been used to establish structure−function relationships of proteins [68]. Raman spectroscopy as well as IR spectroscopy is very useful in biophysics characterization, which provides wealth details when coupled with electrochemistry techniques.

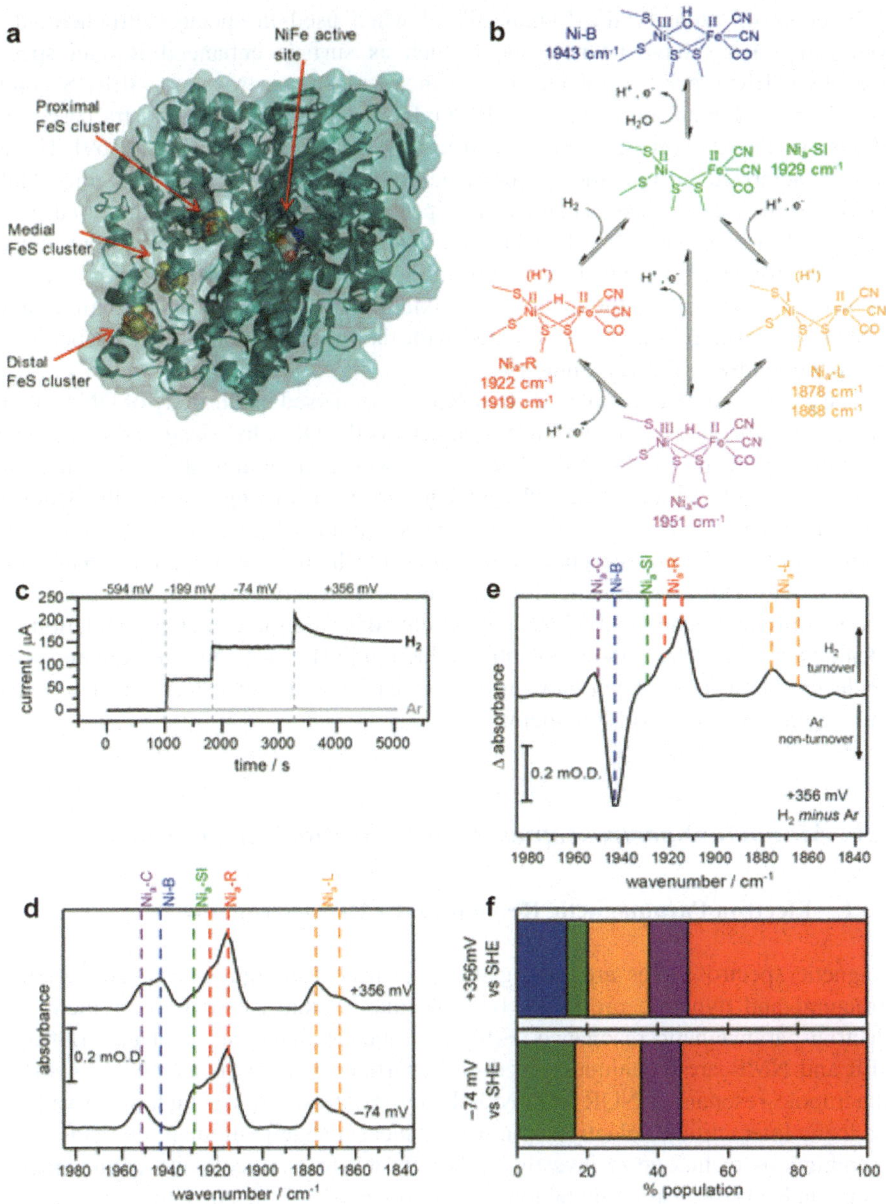

Fig. 4 **a** Structure of [NiFe] hydrogenase I from *E. coli*, PDB 3UQY. **b** Catalytic cycle: states are color-coded to match IR spectra shown later; νCO band positions refer to Hyd1, pH 6.0. **d** Electrocatalytic current response, compared to the corresponding current in the absence of turnover; **e** IR spectra recorded at −74 and +356 mV during steady-state H_2 oxidation; **f** H_2 turnover minus Ar nonturnover difference spectrum at +356 mV; **g** column charts showing the steady-state distribution of active site states at −74 and +356 mV during H_2 oxidation. The protein film IR electrochemistry of *E. coli* Hyd1 was carried out under turnover conditions (1 bar H_2). Reprinted with permission from Ref. [57]. Copyright © 2019, American Chemical Society

In order to intensify the Raman signal when used in spectroeletrochemistry some approaches have been adopted, such as surface enhanced Raman spectroscopy (SERS), surface enhanced resonance Raman spectroscopy (SERRS), and tip-enhanced Raman spectroscopy (TERS). In situ spectroeletrochemistry based on SERRS was used to investigate the electrocatalysis of microbial biofilms [69]. There was characterized the outer membrane cytochromes in a catalytically active microbial biofilm by measuring electrochemical and spectroscopic properties of microbial cells embedded in their natural biofilm habitat. The experiments were carried out under strict electrochemical control, revealing the redox, coordination and spin states of the heme iron as well as the nature of its axial ligand, thereby providing important structural information that corroborated with the interpretation of electrochemical data obtained by cyclic voltammetry.

The electrochemical coupled with SERS has been used in the study of DNA, their bases and nucleotides. An aptasensor that detects the DNA hybrization directly was evaluated by in situ SERS [70]. There was observed an unique spectral profile for each of the four bases and nucleotides, which were easily distinguished. Other studies using in situ SERS applied in the DNA bases and nucleotides were reported, such as the study of adenine and guanine oxidation mechanism [71] and the monitoring of DNA dehybridization kinetics for single nucleotide polymorphism detection [72]. The in situ electrochemical TERS helped elucidate the molecular reorientation of adenine on gold surface under potential induction [73]. Therefore, the Raman spectroelectrochemistry has been showing an outstanding tool to investigate nucleic acids and the mechanisms involving them.

3.3 Magnetic Spectroscopies Applied in Bioelectrochemistry

3.3.1 Electron Paramagnetic Resonance (EPR) Spectroscopy

Magnetic spectroscopies are among the most used techniques for characterizing structural and dynamic properties of molecular systems, examples of them are electron paramagnetic resonance (EPR) and nuclear magnetic resonance (NMR). EPR and NMR are techniques based on hyperfine interactions as well as nuclear quadrupole resonance (NQR) and Mössbauer spectroscopies, which are based on magnetic interaction of electron spin with nuclear magnetic momentum. EPR is an unambiguous technique to investigate directly the free radicals and paramagnetic compounds viz., transition metal ions, triplet states, etc. [74]. Thus, this technique is very useful in the study of bioinorganic structures (as metalloproteins) [75, 76] and biomechanisms involving free radicals [77, 78].

EPR spectroscopy is a powerful tool for the investigation of electron transfer reactions, since one-electron transfer must involve paramagnetic species. Then, the electrochemistry can be used to promote the electron transfer giving access to less stable radicals [79]. Operando EPR in bioelectrochemistry has been used to study the mechanisms of redox reaction in biomolecules [80, 81]. The protein film electrochemical combined with EPR (PFE-EPR) provided an impressive potential control of a redox-active protein in the absence of any redox mediators for EPR spectroscopic investigations [80]. There was studied a Cu–Zn superoxide dismutase modified meso-ITO electrode. By sweeping the potential from ~0.8 to ~−0.4 V versus the standard hydrogen electrode, the EPR signal intensity decreased with the reduction of Cu^{2+} to Cu^{+} at the active redox center of the enzyme.

Operando EPR has been used also for elucidating the electron transfer mechanism of coenzymes [81]. This technique was used investigate to the NADH/NAD^{+} (nicotinamide adenine dinucleotide) redox reaction and introduce how to probe free electrons on a carbon electrode surface and correlate them with the electrocatalytic mechanism. Figure 5a shows the operando EPR electrochemical cell that was designed for use with aqueous electrolyte. This cell consisted in a capillary system with three electrodes comprising flexible carbon fibers (FCF) pristine and functionalized as working electrode (WE), platinum as the counter electrode, and saturated Ag/AgCl as the reference electrode. Figure 5b (absence of NADH) and c (in presence of NADH) present the X-band EPR spectra recorded simultaneously with chronoamperometric measurements after 50 s, when the steady-state current was achieved. These experiments were carried out by using three different kinds of WE, pristine FCF, FCF oxidized (FCF-O), and FCF functionalized with anthraquinone (FCF-AQ). The presence of quinone functionalities causes an increase in the number of unpaired

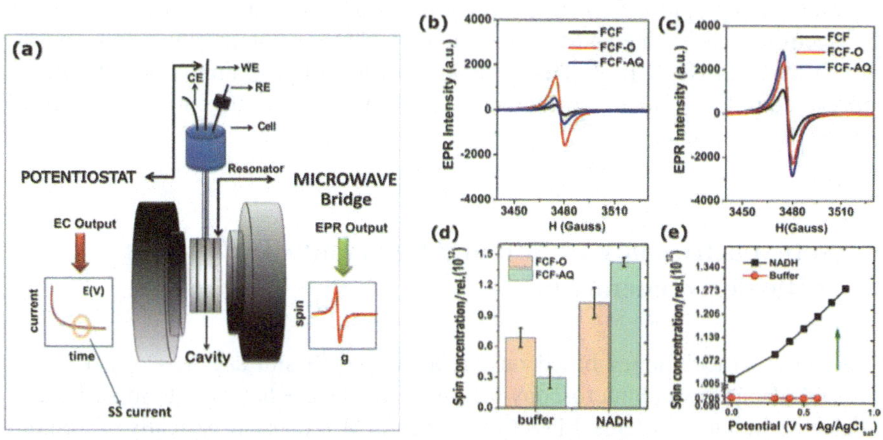

Fig. 5 **a** Operando EPR cell design for applications in electrochemistry. X-band EPR spectra of pristine FCF (black line), FCF-O (red line), and FCF-AQ (blue line) in a phosphate buffer solution (0.1 mol L^{-1}, pH 7.5) in **b** the absence of NADH and **c** the presence of NADH (1.0 mmol L^{-1}) at 0.60 V. **d** Spin concentration obtained with FCF-O (orange bars) and FCF-AQ (green bars) in a phosphate buffer solution (0.1 mol L^{-1}, pH 7.5) in the absence of NADH and in the presence of NADH (1.0 ^{mmol} L^{-1}). **e** Plot of the relative spin concentration versus the applied potential for FCF-O in the absence of NADH (red dots) and in the presence of NADH (1.0 mmol L^{-1}) (black dots). All measurements were carried out at 273 K. Reprinted with permission of Ref. [81]. Copyright © 2019, American Chemical Society

electrons on the electrode surface, thus it is possible to observe higher EPR signals for FCF-O and FCF-AQ. The presence of NADH in the electrolyte media provided a concentration of unpaired electrons increases for all electrodes, which is observed by intensification of the EPR signal. Figure 5d shows the difference of spin concentration in absence and in presence of NADH for the working electrodes based on FCF-O and FCF-AQ. The spin concentration was measured in function of potential using FCF-O in absence and in presence of NADH as presented in Fig. 5e.

3.3.2 Nuclear Magnetic Resonance (NMR) Spectroscopy

Modern molecular biology makes heavy use of NMR to have knowledge of the structure and dynamics of biomacromolecules such as proteins, nucleic acids, or oligosaccharides. While in situ EPR spectroelectrochemistry is the method of choice to detect paramagnetic structures and to characterize their electronic state via the g-values and hyperfine interactions in redox reactions, in situ NMR spectroelectrochemistry provides a powerful method for the investigation of structural or electronic changes of diamagnetic molecules in any electrochemical reaction mechanism [15].

Although NMR spectroelectrochemistry appears as a promising tool for investigation of biomolecules, there are few scientific articles in bioelectrochemistry. One example is the study of the redox reaction of ascorbic acid that acts as coenzyme was studied by using in situ ^{1}H NMR [82]. Other scientific fields have developed interesting approaches for studying redox systems, such as the evaluation of the electron transfer rate and the electrolyte decomposition in redox flow batteries by using in situ NMR [83]. Thus, this kind of technique exhibits potentialities to apply in complexes bioelectrochemical systems that containing biomacromolecules as proteins and nucleic acids.

4 Non-spectroscopy In Situ and Operando Techniques in Bioelectrochemistry

In addition to techniques based on operando and in situ spectroscopy, there are mass spectrometry and microscopy merged with electrochemistry to investigate the bioelectrochemical systems. Mass spectrometry (MS) consists basically of weighing ions in the gas phase, i.e., the measurement of the ratio mass/charge (m/z) of ionic fragments and/or molecular ion. MS measure is an important tool to analyze biomolecular structures qualitatively and quantitively, for example, MS is an important technique in proteomics field [84]. The other non-spectroscopy techniques used in bioelectrochemistry is the atomic force microscopy (AFM) [85] and scanning tunneling microscopy (STM) [86–89], which are used to sense a probe-to-surface atom interaction.

In bioelectrochemistry, the operando MS was used to evaluate the enzymatic kinetics of alcohol dehydrogenase (ADH) that is a NAD-dependent enzyme [90]. The ADH promotes the redox reaction of short-chain alcohols in the presence of the coenzyme NAD^+. For instance, the ethanol can be oxidized in acetaldehyde while the NAD^+ is reduced in NADH. These products can be measured by operando MS, since the acetaldehyde is a very volatile compound and the NADH is oxidized easily, then they are measurable by MS and electrochemistry techniques, respectively. Acetaldehyde production was detected by monitoring of the m/z 29 that corresponds to the main fragment (COH^+) of this product, the m/z 22 was monitored as blank, since this ratio corresponds to doubly charged carbon dioxide that is not produced in the ethanol oxidation by ADH. The NADH production was monitored by current measurement under potential control by using a potentiostat. There was used FCF functionalized with quinone-like groups that acts as electrocatalysts in the NADH electro-oxidation [91]. The setup of the operando MS (also called differential electrochemical mass spectrometry, DEMS), is presented in Fig. 6a. This approach consisted in a conventional electrochemical cell coupled with mass spectrometer by using very hydrophobic interface that allowed only the passage of volatile compounds. Figure 6b, c show the measure of faradaic and ionic currents by the time, respectively. The faradaic chronoamperograms were obtained with the polarization of ADH bioelectrode at 0.6 versus $Ag/AgCl/Cl^-{}_{sat}$, where is observed the increase of current with ethanol increments. The ionic chronoamperograms were obtained by monitoring the m/z 29 and 22 and ADH bioelectrode was polarized at 0.0 and 0.6 V versus $Ag/AgCl/Cl^-{}_{sat}$ at different times. The black line corresponds to the ionic current of m/z 22 and was used as the background, since no CO_2 production was expected. Other lines are the ionic currents measured for m/z 29, which showed an increase when the bioelectrode was polarized at 0.6 V versus $Ag/AgCl/Cl^-{}_{sat}$ and the intensity was higher by increasing ethanol concentration in the electrolyte media. From faradaic and ionic currents values at steady-state, there was possible to establish the enzymatic kinetic parameters by using Michaelis_Menten model, as K_m/K_{cat}.

The in situ electrochemical microscopies have been used to probe surface under potential or current control of the analyzed surface. In bioelectrochemistry, there is the possibility to study redox proteins, entire cells, and nucleic acids by using in situ electrochemical microscopies. The electrochemical atomic microscopy (AFM-SECM) operated in molecule touching mode and combined with redox immunomarking, enables the in situ mapping of the distribution of proteins on individual virus particles and provided the localization of individual viral proteins possible [85]. The in situ STM allowed the evaluation of metalloproteins assembled on gold surfaces under potential control [86, 87], this kind of approach allows a single-molecule measure [87]. Therefore, the in situ electrochemical microscopies have showing as an excellent tool to assess details of biomolecular structure and how the electron transfer occurs in bioelectrochemical systems.

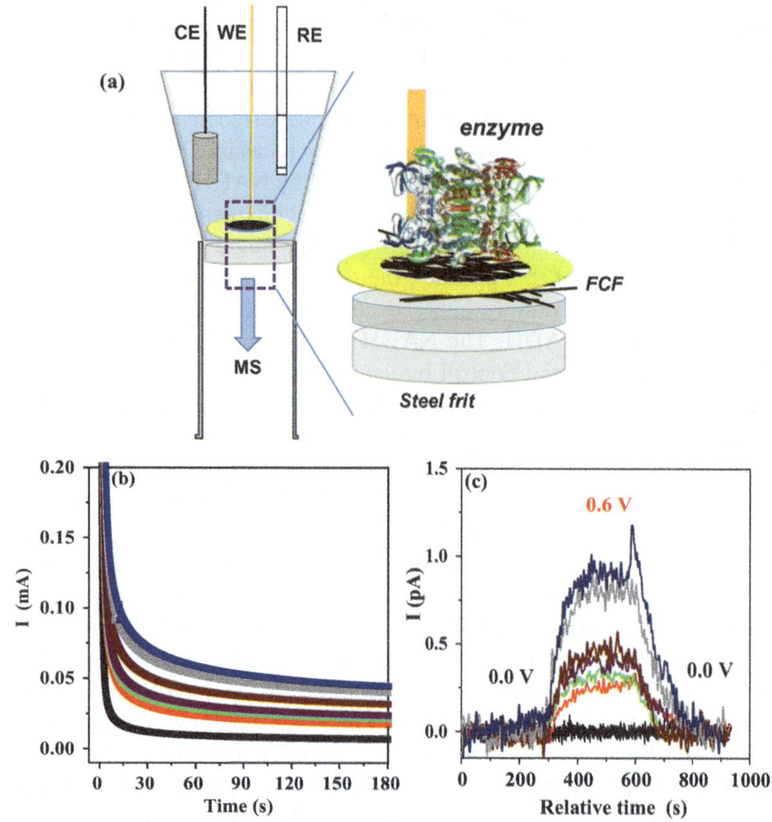

Fig. 6 **a** DEMS setup showing the electrochemical cell, where CE stands for the counter-electrode, WE stands for the working electrode, the black region denotes flexible carbon fiber with immobilized ADH, the yellow region is a Au electrical connection, and RE is the reference electrode ($Ag/AgCl/Cl^{-}{}_{sat}$). The interface comprises a polytetrafluoroethylene membrane covering a steel frit. **b** Chronoamperograms at 0.6 V versus $Ag/AgCl/Cl^{-}{}_{sat}$ of FCF-ADH recorded in the DEMS experiment for ethanol concentrations of up to 0.8 mol L^{-1}. **c** Mass intensity peak monitoring at $m/z = 29$ and 22 upon the application of overpotentials of 0.6 V and 0.0 V (vs. $Ag/AgCl/Cl^{-}{}_{sat}$), each line represents an MS chronoamperogram. Adapted from Ref. [90] with permission from the Royal Society of Chemistry

5 Overview

In situ and operando techniques in bioelectrochemistry appears as powerful tools to investigate biomolecules and biosystems deeply. Since it is possible to control the electron transfer by modulating the potential on bioelectrode interface and monitors structural changes in the bioelectrochemical systems, which arises from of specifically reagents/products or conformational changes on their structures. Moreover, some physicochemical parameters can be changed during the measurements, so it is possible to discover thermodynamics and kinetics parameters. There is a plenty of possibilities in the development and the using of the in situ and operando techniques in bioelectrochemistry. For instance, the in situ NMR is still unexplored for complexes biomacromolecules, although the NMR standalone being one of the most powerful tools in the structure elucidation and dynamics evaluation for biomolecules. Therefore, the use and the development of these approaches are the cutting-edge in the bioelectrochemistry field and can bring in-depth insights of biomolecular mechanisms.

References

1. de Souza, J.C.P., Macedo, L.J., Hassan, A., Sedenho, G.C., Modenez, I.A., Crespilho, F.N.: In situ and operando techniques for investigating electron transfer in biological systems. ChemElectroChem **8**(3), 431–446 (2021)
2. Pereira, A.R., Sedenho, G.C., de Souza, J.C.P., Crespilho, F.N.: Advances in enzyme bioelectrochemistry. An. Acad. Bras. Ciênc. **90**, 825–857 (2018)
3. Kornienko, N., Ly, K.H., Robinson, W.E., Heidary, N., Zhang, J.Z., Reisner, E.: Advancing techniques for investigating the enzyme–electrode interface. Acc. Chem. Res. **52**(5), 1439–1448 (2019)
4. Lopez-Lorente, A.I., Kranz, C.: Recent advances in biomolecular vibrational spectroelectrochemistry. Curr. Opin. Electrochem. **5**(1), 106–113 (2017)
5. Murgida, D.H.: In situ spectroelectrochemical investigations of electrode-confined electron-transferring proteins and redox enzymes. ACS Omega **6**(5), 3435–3446 (2021)
6. Ash, P.A., Vincent, K.A.: Spectroscopic analysis of immobilised redox enzymes under direct electrochemical control. Chem. Commun. **48**(10), 1400–1409 (2012)
7. Hexter, S.V., Esterle, T.F., Armstrong, F.A.: A unified model for surface electrocatalysis based on observations with enzymes. Phys. Chem. Chem. Phys. **16**(24), 11822–11833 (2014)
8. Luz, R.A., Pereira, A.R., de Souza, J.C.P., Sales, F.C.P.F., Crespilho, F.N.: Enzyme biofuel cells: thermodynamics, kinetics and challenges in applicability. ChemElectroChem **1**(11), 1751–1777 (2014)
9. Weckhuysen, B.M.: In-Situ Spectroscopy of Catalysts. American Scientific Publishers Stevenson Ranch, CA (2004)
10. Ren, X., Wang, E., Skitnevskaya, A.D., Trofimov, A.B., Gokhberg, K., Dorn, A.: Experimental evidence for ultrafast intermolecular relaxation processes in hydrated biomolecules. Nat. Phys. **14**(10), 1062–1066 (2018)
11. León, L., Mozo, J.D.: Designing spectroelectrochemical cells: a review. TrAC, Trends Anal. Chem. **102**, 147–169 (2018)
12. Bott-Neto, J.L., et al.: Versatile spectroelectrochemical cell for in situ experiments: development, applications, and electrochemical behavior. ChemElectroChem **7**(21), 4306–4313 (2020)
13. Garoz-Ruiz, J., Perales-Rondon, J.V., Heras, A., Colina, A.: Spectroelectrochemical sensing: current trends and challenges. Electroanalysis **31**(7), 1254–1278 (2019)
14. Zhai, Y., Zhu, Z., Zhou, S., Zhu, C., Dong, S.: Recent advances in spectroelectrochemistry. Nanoscale **10**(7), 3089–3111 (2018)
15. Klod, S., Dunsch, L.: A combination of in situ ESR and in situ NMR spectroelectrochemistry for mechanistic studies of electrode reactions: the case of p-benzoquinone. Magn. Reson. Chem. **49**(11), 725–729 (2011)
16. Ramírez, P., Mano, N., Andreu, R., Ruzgas, T., Heller, A., Gorton, L., Shleev, S.: Direct electron transfer from graphite and functionalized gold electrodes to T1 and T2/T3 copper centers of bilirubin oxidase. Biochim. Biophys. Acta (BBA)-Bioenerg. **1777**(10), 1364–1369 (2008)

17. Dau, H., Liebisch, P., Haumann, M.: X-ray absorption spectroscopy to analyze nuclear geometry and electronic structure of biological metal centers—potential and questions examined with special focus on the tetra-nuclear manganese complex of oxygenic photosynthesis. Anal. Bioanal. Chem. **376**(5), 562–583 (2003)
18. Clark, K., Penner-Hahn, J.E., Whittaker, M.M., Whittaker, J.W.: Oxidation-state assignments for galactose oxidase complexes from X-ray absorption spectroscopy. Evidence for copper (II) in the active enzyme. J. Am. Chem. Soc. **112**(17), 6433–6434 (1990)
19. Wang, H., Patil, D.S., Gu, W., Jacquamet, L., Friedrich, S., Funk, T., Cramer, S.P.: L-edge X-ray absorption spectroscopy of some Ni enzymes: probe of Ni electronic structure. J. Electron Spectrosc. Relat. Phenom. **114**, 855–863 (2001)
20. Tierney, D.L., Fee, J.A., Ludwig, M.L., Penner-Hahn, J.E.: X-ray absorption spectroscopy of the iron site in escherichia coli Fe (III) superoxide dismutase. Biochemistry **34**(5), 1661–1668 (1995)
21. Fabbri, E., Abbott, D.F., Nachtegaal, M., Schmidt, T.J.: Operando X-ray absorption spectroscopy: a powerful tool toward water splitting catalyst development. Curr. Opin. Electrochem. **5**(1), 20–26 (2017)
22. Ortega, R., Carmona, A., Llorens, I., Solari, P.L.: X-ray absorption spectroscopy of biological samples. A tutorial. J. Anal. At. Spectrom. **27**(12), 2054–2065 (2012)
23. Arcovito, A., et al.: X-ray structure analysis of a metalloprotein with enhanced active-site resolution using in situ x-ray absorption near edge structure spectroscopy. Proc. Natl. Acad. Sci. **104**(15), 6211–6216 (2007)
24. Strange, R.W., Feiters, M.C.: Biological X-ray absorption spectroscopy (BioXAS): a valuable tool for the study of trace elements in the life sciences. Curr. Opin. Struct. Biol. **18**(5), 609–616 (2008)
25. del Barrio, M., Fourmond, V.: Redox (In) activations of metalloenzymes: a protein film voltammetry approach. ChemElectroChem **6**(19), 4949–4962 (2019)
26. Macedo, L.J., Hassan, A., Sedenho, G.C., Crespilho, F.N.: Assessing electron transfer reactions and catalysis in multicopper oxidases with operando X-ray absorption spectroscopy. Nat. Commun. **11**(1), 1–7 (2020)
27. Förster, H.: UV/vis spectroscopy. In: Characterization I, pp. 337–426. Springer, Berlin, Heidelberg (2004)
28. Chen, Z., et al.: UV-vis spectroscopy. In: Photoelectrochemical Water Splitting, pp. 49–62. Springer, New York, NY (2013)
29. Vogt, S., Schneider, M., Schäfer-Eberwein, H., Nöll, G.: Determination of the pH dependent redox potential of glucose oxidase by spectroelectrochemistry. Anal. Chem. **86**(15), 7530–7535 (2014)
30. Wu, S.P., Bellei, M., Mansy, S.S., Battistuzzi, G., Sola, M., Cowan, J.A.: Redox chemistry of the Schizosaccharomyces pombe ferredoxin electron-transfer domain and influence of Cys to Ser substitutions. J. Inorg. Biochem. **105**(6), 806–811 (2011)
31. Dong, S., Niu, J., Cotton, T.M.: [28] Ultraviolet/visible spectroelectrochemistry of redox proteins. Methods Enzymol. **246**, 701–732 (1995)
32. Lakowicz, J.R. (ed.): Principles of Fluorescence Spectroscopy. Springer Science & Business Media, New York (2013)
33. Stryer, L.: Fluorescence spectroscopy of proteins: fluorescent probes provide insight into the structure, interactions, and dynamics of proteins. Science **162**(3853), 526–533 (1968)
34. Ladokhin, A.S.: Fluorescence spectroscopy in peptide and protein analysis. In: Encyclopedia of Analytical Chemistry: Applications, Theory and Instrumentation (2006)
35. Lakowicz, J.R.: Fluorescence spectroscopic investigations of the dynamic properties of proteins, membranes and nucleic acids. J. Biochem. Biophys. Methods **2**(1–2), 91–119 (1980)
36. Schurr, J.M., Fujimoto, B.S., Wu, P., Song, L.: Fluorescence studies of nucleic acids: dynamics, rigidities, and structures. In: Topics in Fluorescence Spectroscopy, pp. 137–229. Springer, Boston, MA (2002)
37. Ma, L., Yang, F., Zheng, J.: Application of fluorescence resonance energy transfer in protein studies. J. Mol. Struct. **1077**, 87–100 (2014)

38. Davis, J.J., Burgess, H., Zauner, G., Kuznetsova, S., Salverda, J., Aartsma, T., Canters, G.W.: Monitoring interfacial bioelectrochemistry using a FRET switch. J. Phys. Chem. B **110**(41), 20649–20654 (2006)
39. Salverda, J.M., et al.: Fluorescent cyclic voltammetry of immobilized azurin: direct observation of thermodynamic and kinetic heterogeneity. Angew. Chem. Int. Ed. **49**(33), 5776–5779 (2010)
40. Patil, A.V., Davis, J.J.: Visualizing and tuning thermodynamic dispersion in metalloprotein monolayers. J. Am. Chem. Soc. **132**(47), 16938–16944 (2010)
41. Patil, A.V., Davis, J.J.: Molecular scale bioelectrochemistry. Coord. Chem. Rev. **255**(17–18), 1970–1980 (2011)
42. Akkilic, N., Kamran, M., Stan, R., Sanghamitra, N.J.: Voltage-controlled fluorescence switching of a single redox protein. Biosens. Bioelectron. **67**, 747–751 (2015)
43. Zhou, L., Arugula, M.A., Chin, B.A., Simonian, A.L.: Simultaneous surface plasmon resonance/fluorescence spectroelectrochemical in situ monitoring of dynamic changes on functional interfaces: a study of the electrochemical proximity assay model system. ACS Appl. Mater. Interfaces **10**(48), 41763–41772 (2018)
44. Bizzotto, D.: In situ spectroelectrochemical fluorescence microscopy for studying electrodes modified by molecular adsorbates. Curr. Opin. Electrochem. **7**, 161–171 (2018)
45. Casanova-Moreno, J.R., Bizzotto, D.: Frequency response analysis of potential-modulated orientation changes of a DNA self-assembled layer using spatially resolved fluorescence measurements. Electrochim. Acta **162**, 62–71 (2015)
46. Yu, Z.L., Casanova-Moreno, J., Guryanov, I., Maran, F., Bizzotto, D.: Influence of surface structure on single or mixed component self-assembled monolayers via in situ spectroelectrochemical fluorescence imaging of the complete stereographic triangle on a single crystal Au bead electrode. J. Am. Chem. Soc. **137**(1), 276–288 (2015)
47. Chen, W., Chen, J.J., Lu, R., Qian, C., Li, W.W., Yu, H.Q.: Redox reaction characteristics of riboflavin: a fluorescence spectroelectrochemical analysis and density functional theory calculation. Bioelectrochemistry **98**, 103–108 (2014)
48. Zaino, L.P., Grismer, D.A., Han, D., Crouch, G.M., Bohn, P.W.: Single occupancy spectroelectrochemistry of freely diffusing flavin mononucleotide in zero-dimensional nanophotonic structures. Faraday Discuss. **184**, 101–115 (2015)
49. Kelly, S.M., Jess, T.J., Price, N.C.: How to study proteins by circular dichroism. Biochim. Biophys. Acta (BBA)-Proteins Proteomics **1751**(2), 119–139 (2005)
50. del Villar-Guerra, R., Gray, R.D., Chaires, J.B.: Characterization of quadruplex DNA structure by circular dichroism. Curr. Protoc. Nucleic Acid Chem. **68**(1), 17–18 (2017)
51. Sutherland, J.C., Holmquist, B.: Magnetic circular dichroism of biological molecules. Annu. Rev. Biophys. Bioeng. **9**(1), 293–326 (1980)
52. Karon, K., Łapkowski, M., Dobrowolski, J.C.: ECD spectroelectrochemistry: a review. Spectrochim. Acta Part A: Mol. Biomol. Spectrosc. **250**, 119349 (2021)
53. Zhu, Y., Cheng, G., Dong, S.: Structural electrochemical study of hemoglobin by in situ circular dichroism thin layer spectroelectrochemistry. Biophys. Chem. **97**(2–3), 129–138 (2002)
54. Syed, S.N., Schulze, H., Macdonald, D., Crain, J., Mount, A.R., Bachmann, T.T.: Cyclic denaturation and renaturation of double-stranded DNA by redox-state switching of DNA intercalators. J. Am. Chem. Soc. **135**(14), 5399–5407 (2013)
55. Marritt, S.J., van Wonderen, J.H., Cheesman, M.R., Butt, J.N.: Magnetic circular dichroism of hemoproteins with in situ control of electrochemical potential: "MOTTLE." Anal. Biochem. **359**(1), 79–83 (2006)
56. Siebert, F., Hildebrandt, P.: Vibrational Spectroscopy in Life Science. Wiley, Weinheim (2008)
57. Ash, P.A., Kendall-Price, S.E., Vincent, K.A.: Unifying activity, structure, and spectroscopy of [NiFe] hydrogenases: combining techniques to clarify mechanistic understanding. Acc. Chem. Res. **52**(11), 3120–3131 (2019)
58. Ash, P.A., Hidalgo, R., Vincent, K.A.: Protein film infrared electrochemistry demonstrated for study of H_2 oxidation by a [NiFe] hydrogenase. J. Vis. Exp. JoVE (130) (2017)
59. Ash, P.A., Hidalgo, R., Vincent, K.A.: Proton transfer in the catalytic cycle of [NiFe] hydrogenases: insight from vibrational spectroscopy. ACS Catal. **7**(4), 2471–2485 (2017)

60. Hitaishi, V.P., et al.: Electrostatic-driven activity, loading, dynamics, and stability of a redox enzyme on functionalized-gold electrodes for bioelectrocatalysis. ACS Catal. **8**(12), 12004–12014 (2018)
61. Liu, X., et al.: In Situ spectroelectrochemical characterization reveals cytochrome-mediated electric syntrophy in geobacter coculture. Environ. Sci. Technol. **55**(14), 10142–10151 (2021)
62. Fallah, M.A., Stanglmair, C., Pacholski, C., Hauser, K.: Devising self-assembled-monolayers for surface-enhanced infrared spectroscopy of pH-driven poly-L-lysine conformational changes. Langmuir **32**(29), 7356–7364 (2016)
63. Zeng, L., Wu, L., Liu, L., Jiang, X.: Analyzing structural properties of heterogeneous cardiolipin-bound cytochrome c and their regulation by surface-enhanced infrared absorption spectroscopy. Anal. Chem. **88**(23), 11727–11733 (2016)
64. López-Lorente, Á.I., Izquierdo, J., Kranz, C., Mizaikoff, B.: Boron-doped diamond modified with gold nanoparticles for the characterization of bovine serum albumin protein. Vib. Spectrosc. **91**, 147–156 (2017)
65. Paengnakorn, P., et al.: Infrared spectroscopy of the nitrogenase MoFe protein under electrochemical control: potential-triggered CO binding. Chem. Sci. **8**(2), 1500–1505 (2017)
66. Hidalgo, R., Ash, P.A., Healy, A.J., Vincent, K.A.: Infrared spectroscopy during electrocatalytic turnover reveals the Ni-L active site state during H_2 oxidation by a NiFe hydrogenase. Angew. Chem. Int. Ed. **54**(24), 7110–7113 (2015)
67. Smith, E., Dent, G.: Modern Raman Spectroscopy: A Practical Approach. Wiley (2019)
68. Rygula, A., Majzner, K., Marzec, K.M., Kaczor, A., Pilarczyk, M., Baranska, M.: Raman spectroscopy of proteins: a review. J. Raman Spectrosc. **44**(8), 1061–1076 (2013)
69. Millo, D., Harnisch, F., Patil, S.A., Ly, H.K., Schröder, U., Hildebrandt, P.: In situ spectroelectrochemical investigation of electrocatalytic microbial biofilms by surface-enhanced resonance Raman spectroscopy. Angew. Chem. Int. Ed. **50**(11), 2625–2627 (2011)
70. Karaballi, R.A., Nel, A., Krishnan, S., Blackburn, J., Brosseau, C.L.: Development of an electrochemical surface-enhanced Raman spectroscopy (EC-SERS) aptasensor for direct detection of DNA hybridization. Phys. Chem. Chem. Phys. **17**(33), 21356–21363 (2015)
71. Ibañez, D., Santidrian, A., Heras, A., Kalbáč, M., Colina, A.: Study of adenine and guanine oxidation mechanism by surface-enhanced Raman spectroelectrochemistry. J. Phys. Chem. C **119**(15), 8191–8198 (2015)
72. Kayran, Y.U., Cinar, N., Jambrec, D., Schuhmann, W.: Monitoring potential-induced DNA dehybridization kinetics for single nucleotide polymorphism detection by using in situ surface enhanced raman scattering. ChemElectroChem **5**(5), 756–760 (2018)
73. Martín Sabanés, N., Ohto, T., Andrienko, D., Nagata, Y., Domke, K.F.: Electrochemical TERS elucidates potential-induced molecular reorientation of adenine/Au (111). Angew. Chem. **129**(33), 9928–9933 (2017)
74. Karunakaran, C., Balamurugan, M.: Electron paramagnetic resonance spectroscopy. In: Karunakaran, C. (ed.) Spin Resonance Spectroscopy: Principles and Applications. Elsevier (2018)
75. Eisermann, J., Seif-Eddine, M., Roessler, M.M.: Insights into metalloproteins and metallodrugs from electron paramagnetic resonance spectroscopy. Curr. Opin. Chem. Biol. **61**, 114–122 (2021)
76. Jasniewski, A., Hu, Y., Ribbe, M.W.: Electron paramagnetic resonance spectroscopy of metalloproteins. In: Metalloproteins, pp. 197–211. Humana Press, New York, NY (2019)
77. Zapp, C., et al.: Mechanoradicals in tensed tendon collagen as a source of oxidative stress. Nat. Commun. **11**(1), 1–8 (2020)
78. Mostert, A.B., Rienecker, S.B., Noble, C., Hanson, G.R., Meredith, P.: The photoreactive free radical in eumelanin. Sci. Adv. **4**(3), eaaq1293 (2018)
79. Murray, P.R., et al.: An in situ electrochemical cell for Q-and W-band EPR spectroscopy. J. Magn. Reson. **213**(1), 206–209 (2011)
80. Abdiaziz, K., Salvadori, E., Sokol, K.P., Reisner, E., Roessler, M.M.: Protein film electrochemical EPR spectroscopy as a technique to investigate redox reactions in biomolecules. Chem. Commun. **55**(60), 8840–8843 (2019)

81. Ali, M.A., Hassan, A., Sedenho, G.C., Gonçalves, R.V., Cardoso, D.R., Crespilho, F.N.: Operando electron paramagnetic resonance for elucidating the electron transfer mechanism of coenzymes. J. Phys. Chem. C **123**(26), 16058–16064 (2019)
82. Da Silva, P.F., et al.: Electrochemical NMR spectroscopy: electrode construction and magnetic sample stirring. Microchem. J. **146**, 658–663 (2019)
83. Zhao, E.W., et al.: Coupled in situ NMR and EPR studies reveal the electron transfer rate and electrolyte decomposition in redox flow batteries. J. Am. Chem. Soc. **143**(4), 1885–1895 (2021)
84. Aebersold, R., Mann, M.: Mass spectrometry-based proteomics. Nature **422**(6928), 198–207 (2003)
85. Nault, L., et al.: Electrochemical atomic force microscopy imaging of redox-immunomarked proteins on native potyviruses: from subparticle to single-protein resolution. ACS Nano **9**(5), 4911–4924 (2015)
86. Zhang, J., Christensen, H.E., Ooi, B.L., Ulstrup, J.: In situ STM imaging and direct electrochemistry of pyrococcus furiosus ferredoxin assembled on thiolate-modified Au (111) surfaces. Langmuir **20**(23), 10200–10207 (2004)
87. Yan, J., et al.: Voltammetry and single-molecule in situ scanning tunnelling microscopy of the redox metalloenzyme human sulfite oxidase. ChemElectroChem **8**(1), 164–171 (2021)
88. Elliott, M.: High-resolution electrochemical STM of redox metalloproteins. Curr. Opin. Electrochem. **4**(1), 152–158 (2017)
89. Zeng, D., Salvatore, P., Karlsen, K.K., Zhang, J., Wengel, J., Ulstrup, J.: Electrochemical intercalator binding to single-and double-strand DNA-and LNA-based molecules on Au (111)-electrode surfaces. J. Electroanal. Chem. **865**, 114138 (2020)
90. de Souza, J.C.P., Silva, W.O., Lima, F.H., Crespilho, F.N.: Enzyme activity evaluation by differential electrochemical mass spectrometry. Chem. Commun. **53**(60), 8400–8402 (2017)
91. Pereira, A.R., de Souza, J.C.P., Gonçalves, A.D., Pagnoncelli, K.C., Crespilho, F.N.: Bioelectrooxidation of ethanol using NAD-dependent alcohol dehydrogenase on oxidized flexible carbon fiber arrays. J. Braz. Chem. Soc. **28**(9), 1698–1707 (2017)

The manufacturer's authorised representative in the EU is Springer Nature Customer Service Centre GmbH, Europaplatz 3, 69115 Heidelberg, Germany. If you have any concerns regarding our products, please contact ProductSafety@springernature.com

Printed and bound by CPI Group (UK) Ltd, Croydon, CR0 4YY
30/09/2026
02225309-0004